MILLIMETER-WAVE DIGITALLY INTENSIVE FREQUENCY GENERATION IN CMOS

MILLIMETER-WAVE DIGITALLY INTENSIVE FREQUENCY GENERATION IN CMOS

WANGHUA WU

ROBERT BOGDAN STASZEWSKI

JOHN R. LONG

AMSTERDAM • BOSTON • HEIDELBERG • LONDON
NEW YORK • OXFORD • PARIS • SAN DIEGO
SAN FRANCISCO • SINGAPORE • SYDNEY • TOKYO
Academic Press is an imprint of Elsevier

Academic Press is an imprint of Elsevier
125, London Wall, EC2Y 5AS.
525 B Street, Suite 1800, San Diego, CA 92101-4495, USA
225 Wyman Street, Waltham, MA 02451, USA
The Boulevard, Langford Lane, Kidlington, Oxford OX5 1GB, UK

ISBN: 978-0-12-802207-8

British Library Cataloguing-in-Publication Data
A catalogue record for this book is available from the British Library.

Library of Congress Cataloging-in-Publication Data
A catalog record for this book is available from the Library of Congress.

For Information on all Academic Press publications visit our website at http://store.elsevier.com/

Typeset by MPS Limited, Chennai, India
www.adi-mps.com

Printed and bound in the United States of America

Publisher: Jonathan Simpson
Acquisition Editor: Tim Pitts
Editorial Project Manager: Charlie Kent
Production Project Manager: Jason Mitchell
Designer: Greg Harris

CONTENTS

PREFACE

Over the past few decades, frequency synthesis based on analog-intensive phase-locked loops (PLLs) has been the most popular technique employed to provide local oscillator signals for the radio frontend. With aggressive scaling and technological advancement in silicon-based process technologies, particularly CMOS, digitally assisted RF systems are fast becoming a commonplace in the low-GHz bands (i.e., below 10 GHz). The key enabler there is digital signal processing employed to improve the overall system performance via calibration, and also to provide reconfigurability and ease of testability. Particularly in the area of RF frequency synthesis, many universities, research institutes, and companies have since demonstrated various all-digital phase-locked loop (ADPLL) implementations, and ADPLLs are now replacing traditional analog PLLs in consumer electronics supporting various wireless standards, for example, 2G/3G cellular, IEEE 802.11 a/b/g/n/ac, and Bluetooth.

As the frequency spectrum becomes increasingly congested in the low-GHz regime, millimeter-wave (mm-wave) frequency bands (i.e., above 30 GHz) are gaining popularity as they offer large bandwidth to support Giga-bit per second wireless communication without the need for complex modulation schemes, thus achieving low error rates and low energy consumption per bit. Up to this date, there have been many published silicon-based mm-wave analog PLLs, but very few ADPLLs operating above 30 GHz are reported. Little material has been written on the ADPLL design challenges at mm-wave frequencies and the design techniques to address them. Moreover, testing and debugging PLLs to correctly identify any design or fabrication problems would be equally challenging due to the closed-loop operation of the PLL.

In this book, we detail these technical challenges, and discuss the design and implementation of a 60-GHz ADPLL in a conventional widely available CMOS process. We further elaborate on calibration techniques that are especially useful at mm-wave to improve the system performance. We also explain the implemented testability features that facilitate design for test and characterization. This book is organized as follows:

- Chapters 1–3 go over the introduction and review of existing literature. Chapter 1 lays out the motivation and challenges in building an ADPLL for the mm-wave regime, while Chapter 2 presents various existing mm-wave frequency synthesizer architectures. Chapter 3 reviews the building blocks of a frequency synthesizer, which are common to both analog and digital implementations.
- Chapters 4–6 deal with the theory, design, and realization of a mm-wave ADPLL. Chapter 4 covers the basic concepts which are needed to understand the design and operation of an ADPLL, and Chapter 5 discusses mm-wave digitally controlled oscillator (DCO) designs and implementations. Chapter 6 addresses the designs of other key circuit blocks, and demonstrates a 60-GHz ADPLL for use in an FMCW transmitter.
- Chapter 7 explains several calibration techniques used to improve the performance of the 60-GHz ADPLL, while Chapter 8 describes the measurement challenges of a mm-wave frequency synthesizer, and proposes build-in self-test and self-characterization techniques.

The work presented in this book is a culmination of several years of research. We would like to thank and acknowledge the discussions and help we received from past and present colleagues at the Department of Electronics of Delft University of Technology in The Netherlands. We also thank the staff at Elsevier for their support.

Wanghua Wu
Robert Bogdan Staszewski
John R. Long
April 2015

LIST OF ABBREVIATIONS

ADC Analog-to-digital converter
ADPLL All-digital phase-locked loop
AM Amplitude modulation
BiCMOS Bipolar and CMOS
BISC Built-in self-characterization
BIST Build-in self-test
CB Coarse-tuning bank
CKM Modulation clock
CKR Reference clock retimed by oscillator clock
CKV Oscillator (variable) output clock
CLK Clock
CML Current-mode-logic
CMOS Complementary metal-oxide-semiconductor
CT Center-tap
DAC Digital-to-analog converter
DCO Digitally controlled oscillator
DDFS Direct digital frequency synthesizer
DFC Design for characterization
DFT Design for test
DNL Differential nonlinearity
DSP Digital signal processing
EM Electromagnetic
ESD Electrostatic discharge
EVM Error vector magnitude
FB Fine-tuning bank
FCC Federal Communications Commission
FCW Frequency command word
FET Field-effect transistor
FM Frequency modulation
FMCW Frequency-modulated continuous-wave
FREF Frequency reference
Gb/s Gigabit per second
GRO Gated ring oscillator
GSM Global system for mobile (communications)
GUI Graphical user interface
HVAC Heating, ventilating, and air conditioning
IC Integrated circuit
IEEE Institute of Electrical and Electronics Engineers
IF Intermediate frequency
IIR Infinite impulse response
ILFD Injection-locked frequency divider
INL Integral nonlinearity
IO Input/output

IR Interconnect resistance
ISM Industrial, scientific and medical
LF Loop filter
LMS Least mean squares
LO Local oscillator
LPF Low-pass filter
LSB Least significant bit
MB Mid-coarse tuning bank
MIM Metal-insulator-metal
MIMO Multiple-input and multiple-output
mm-wave Millimeter-wave
MoM Metal-oxide-metal
MOS Metal-oxide-semiconductor
MTBF Meantime between failures
nDCO Normalized DCO
NMOS N-type metal-oxide-semiconductor
NTW Normalized tuning word
OTW Oscillator tuning word
PA Power amplifier
PCB Printed circuit board
PFD Phase/frequency detector
PHE Phase error
PHR Phase of frequency reference
PHV Phase of variable oscillator
PI Proportional-integral
PLL Phase-locked loop
PM Phase modulation
PMOS P-type metal-oxide-semiconductor
PN Phase noise
PPF Poly-phase filter
PROM Programmable read-only memory
PVT Process, voltage and temperature
QAM Quadrature amplitude modulation
Q-factor Quality factor
Rx Receiver
RF Radio frequency
RFIC Radio frequency integrated circuit
rms Root-mean-square
RO Ring oscillator
SAFF Sense-amplifier-based flip-flop
SiGe Silicon Germanium
SoC System-on-chip
SPI Serial peripheral interface
SRAM Static random-access memory
TDC Time-to-digital converter
TL Transmission line
TR Tuning range

TSPC True single-phase clocked
Tx Transmitter
UWB Ultra-wideband
VCO Voltage-controlled oscillator
WiGig Wireless Gigabit Alliance
WiMAX Worldwide Interoperability for Microwave Access
WLAN Wireless local-area network
WPAN Wireless personal-area network

CHAPTER 1

Introduction

Contents

Wireless communication has evolved remarkably since Guglielmo Marconi demonstrated the transmission and reception of Morse-coded messages across the Atlantic Ocean in the early twentieth century. Since then, new wireless communication methods and services have been continuously adopted that revolutionize our lives. Today, cellular, mobile, and wireless local-area networks (WLANs), afforded by breakthroughs in semiconductor technologies and their capability of mass production, are in use worldwide. They enable us to share images of our cherished moments with family and friends anywhere, and at anytime. The current trends toward portable wireless devices with ultra-high-speed (e.g., gigabit per second) connectivity will soon allow us to go online via our notebooks, cell phones, and tablets, simultaneously emailing, chatting with friends, web browsing and downloading movies and music in a fraction of the time it takes today. These devices will have to meet aggressive performance specifications in a sufficiently small and low-cost product at low power dissipation. This has prompted frantic research into new radio frequency (RF)-integrated circuits, system architectures, and design approaches.

This book explores the feasibility, advantages, design, and testing of digitally intensive frequency synthesis in the millimeter-wave (mm-wave) frequency range. An all-digital phase-locked loop (ADPLL)-based transmitter

Millimeter-Wave Digitally Intensive Frequency Generation in CMOS.
DOI: http://dx.doi.org/10.1016/B978-0-12-802207-8.00001-0

demonstrator fabricated in a production bulk CMOS process is described, which operates in the 60-GHz band, and achieves fractional frequency generation and wideband frequency modulation (FM). This digitally intensive design has the potential for low cost in volume production. It is also amenable to scaling in future technology nodes as opposed to other analog-intensive implementations. The silicon area and power consumption of such transmitters may be reduced further in future by harnessing the power of digital signal processing (DSP).

1.1 MOTIVATION

To achieve gigabit per second (i.e., Gb/s) transfer rates, Wi-Fi technology (IEEE 802.11ac in the 5-GHz band) [1] has been developed in recent years. Multistation WLAN throughput of at least 1 Gb/s, and a single link throughput of at least 500 Mb/s is specified. It employs RF bandwidths of up to 160 MHz, multiple-input and multiple-output (MIMO) array transmitter/receiver streams (up to 8), multi-user MIMO, and up to 256-QAM (quadrature amplitude modulation) schemes in order to achieve that level of performance. The mm-wave frequency bands, by contrast, are less crowded than the low-gigahertz radio communication bands and, more attractively, have wider license-free RF bandwidth available (e.g., 7 GHz bandwidth in the 60-GHz band). This will enable the gigabit-per-second short-range communication for consumer multimedia products and support the development of emerging short-range wireless networking in many important areas, for example, commerce, manufacturing, transport, etc., and thus provide significant growth potential in new internet applications in price-sensitive communication markets.

In the following sections, the advantages and challenges of mm-wave transceiver design in CMOS technology will be examined. The focus is on mm-wave frequency synthesis.

1.1.1 Advantages of Millimeter-Wave Radios

The mm-wave frequency band is defined as 30–300 GHz with a wavelength between 1 and 10 mm in the air [2]. There are various aspects of mm-wave bands that make it attractive for short-range applications. One major advantage is the bandwidth available to carry information. To keep operating costs low, regulatory licensed bands should be avoided, thus calling for the exploitation of the unlicensed or the industrial, scientific, and medical (ISM) radio bands. Figure 1.1 plots the available bandwidth

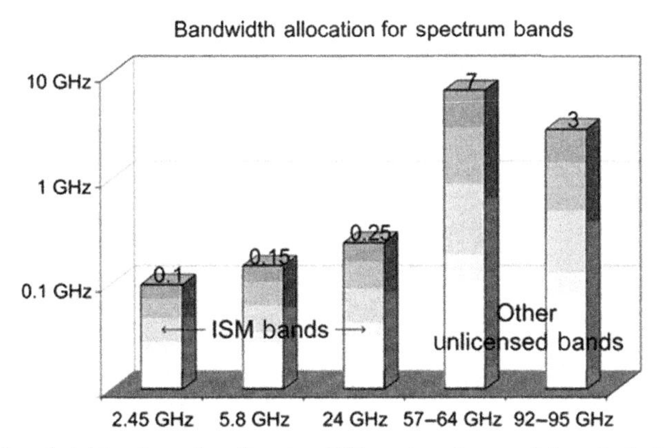

Figure 1.1 Bandwidth allocation for the ISM and unlicensed bands below 100 GHz by the FCC (in the United States) [3].

(indicated in GHz at the top of each column) for ISM and unlicensed bands below 100 GHz in the United States [3]. Below 25 GHz, the RF spectrum is congested due to frequency slots reserved for military, civil, and personal communication services. For reference, most commercial products operate in bands below 10 GHz, for example, the global system for mobile communications operates at 900 and 1,800 MHz (in Europe), and 850 and 1,900 MHz (in the United States), and ultra-wideband (UWB) radios are permitted to operate from 3.1 to 10.6 GHz [4]. Less than 1 GHz of bandwidth in total has been allocated for the license-free ISM bands at 2.45, 5.8, and 24 GHz. On the contrary, there is 7 GHz of bandwidth in the 60-GHz spectrum band allocated for license-free use, which is the largest ever allocated by the Federal Communications Commission (FCC) in the United States below 100 GHz. With such wide bandwidth available, mm-wave wireless links can achieve capacities as high as 7 Gb/s full duplex, which is unlikely to be matched by any of the RF wireless technologies at lower frequencies. The FCC has also recently approved another unlicensed band (92–95 GHz) to meet the growing demand for point-to-point high-bandwidth communication links [5].

For a given antenna size, the beamwidth can be made finer by increasing the frequency. Another benefit of the mm-wave radio is a narrower beam due to the shorter wavelength ($\lambda = c/f_c$, where c is the speed of light and f_c is the carrier frequency), which allows for deployment of multiple, independent links in close proximity. The main limitation of mm-wave radio is the physical range. Due to absorption by atmospheric oxygen and

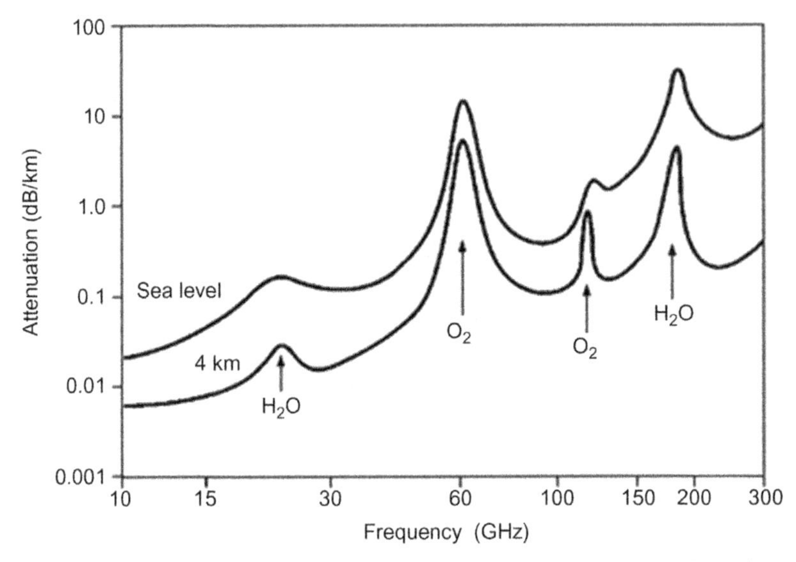

Figure 1.2 Average atmospheric attenuation of radio waves propagating through free space versus frequency [6].

water vapor, signal strength drops off rapidly with distance compared to other bands. Figure 1.2 illustrates the general trend of increasing the attenuation of radio waves with frequency (due only to atmospheric losses; free space path loss is not accounted for) [6]. Atmospheric absorption by oxygen causes more than 15 dB/km of attenuation. The loss of a link budget at 60 GHz is therefore unacceptable for long-distance communication (e.g., >1 km), but can be used to an advantage in short-range indoor communications because the limited range and narrow beamwidths prevent interference between neighboring links. These attributes have led to greatly reduced regulatory burdens for mm-wave communications.

Due to its potential for short-range, gigabit-per-second communications, several standards in the 60-GHz band have been established in recent years. The IEEE 802.15.3c standard was approved in 2009 for wireless personal-area network [7]. A similar standard for Europe (ECMA-387 [8]) was published in 2008. The WirelessHD consortium has released a specification version 1.0a for regulating the transmission of high-definition video in this unlicensed band [9]. Most recently, the IEEE 802.11ad standard (known as WiGig) [10] was adopted in 2013. It provides data rates up to 7 Gb/s, or more than 10× the maximum speed previously supported by the IEEE 802.11 standard. IEEE 802.11ad also adds a "fast session transfer" feature, which enables wireless devices to seamlessly transition between the 60-GHz frequency band and legacy bands at 2.4 and 5 GHz in order to optimize link performance and range criteria.

In addition to the gigabit-per-second communication, the 60-GHz unlicensed band also holds promise for integrating wireless sensors. Frequency-modulated continuous-wave (FMCW) radars may be utilized for presence detection and ranging at 60-GHz applications, where high-frequency resolution is required [11]. This is also the intended application for the realized ADPLL frequency synthesizer that is fully described in this book. As an example of such FMCW application is a gesture recognition system for cars, where the driver gestures (e.g., nodding the head) without taking the eyes off the road when interfacing with applications such as navigation, phone, HVAC (heating, ventilating, and air conditioning) controls, etc. The targeted detection range is from 0.3 to 10 m and the range resolution is below 5 cm. A low-cost implementation of short-range radar systems will enable numerous applications in security, search and rescue, imaging, logistics, quality control, to name just a few.

Figure 1.3 illustrates the operating principle of an FMCW radar transceiver. The carrier signal is modulated as shown in Figure 1.3b, resulting in a signal whose instantaneous frequency varies linearly with time, i.e., a linear chirp [12]. This linear chirp is transmitted toward a target, and the received echo is convolved with a portion of the transmitted signal to determine the round-trip propagation time, τ. In an FMCW radar, the achievable range resolution (Δr) is determined by

$$\Delta r = \frac{c}{2 \cdot BW'} \tag{1.1}$$

where c is the speed of light, and BW (Figure 1.3b) is the modulation range of the transmit signal [12]. When the full 7 GHz of bandwidth at 60 GHz is utilized, a range resolution as fine as 2 cm can be achieved.

1.1.2 Deep-Submicron CMOS

Silicon technologies (e.g., CMOS and SiGe-BiCMOS) are mainstream integrated circuit (IC) processes driven largely by mass production of ICs used in digital computers (e.g., desktops and notebooks) and other electronic devices (e.g., cell phones, game consoles, and tablets). The demand for a higher integration level and lower cost in volume production has driven mm-wave electronics development in silicon CMOS technology. With 65-nm bulk

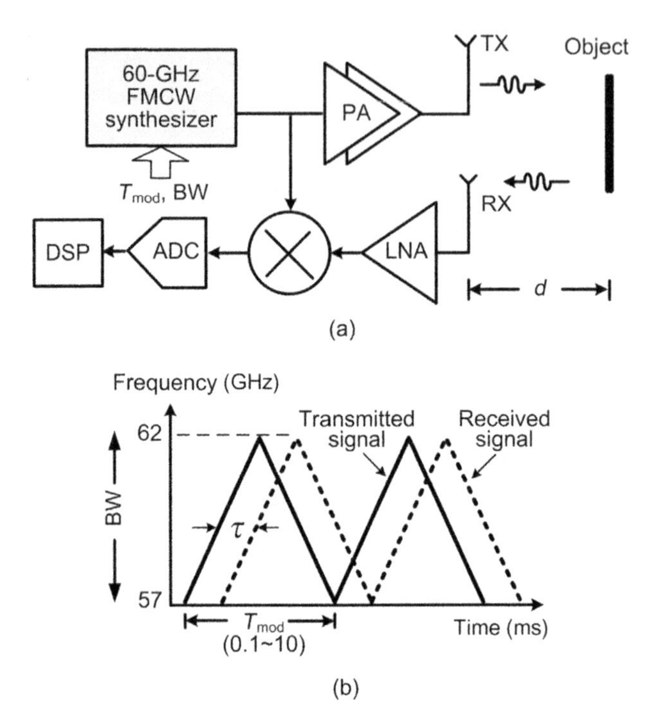

Figure 1.3 (a) Simplified block diagram of an FMCW radar transceiver; (b) transmitted and received linear FMCW signal.

CMOS technologies in production offering peak transit frequency (f_T) and maximum frequency of oscillation (f_{max}) close to 200 GHz (simulated using baseline transistors) [13], several experimental 60-GHz transceivers have been reported that achieve data rates above 4 Gb/s across a 2-m link [14–16]. These prototypes demonstrate the potential to use CMOS for RF/baseband co-integration, and reveal the design challenges and opportunities for improved RF performance (e.g., higher RF output power, lower oscillator phase noise (PN), etc.) and lower power consumption. Nevertheless, III-V processes still have a niche in power amplifiers and antenna switches for mobile phones/base stations, and ultra-high-frequency, high-power electronics for military and space applications.

A better understanding of the properties of deep-submicron CMOS technologies is crucial in order to implement high-performance mm-wave ICs. Several key properties are summarized here, which should be taken into account when deciding on the preferred architectures and design approaches.

- Low supply voltage: compared to III-V and bipolar technology, which rely on a larger supply voltage (3.3 and 2.5 V), deep-submicron CMOS features a nominal supply voltage of ~1 V.
- Due to the aforementioned low supply voltage of deep-submicron CMOS and relatively high threshold voltage (0.5 V and often higher, due to the body effect), the available voltage headroom is quite small. Thus, the margin between technology performance and design requirements appears larger in the time domain than in the voltage domain [17].
- Excellent switching characteristics of MOS transistors—both rise and fall times are on the order of tens of picoseconds or less for deep-submicron CMOS technologies.
- Rapid pace of process scaling—each new digital CMOS process node occurs roughly every 18 months, resulting in an increase in the digital gate density by a factor of 2 (known as Moore's law [18]).
- Multiple metal layers are commonly available for interconnection in large-scale digital circuitry, which also provide high-density metal-oxide-metal capacitors.

1.1.3 Digitally Intensive Approach

Due to the aforementioned properties of deep-submicron CMOS technologies, especially the strength in circuit speed and density, digitally assisted and digitally intensive RF systems are becoming attractive for mm-wave transceiver ICs. When the designed RF system employs digital logic and signal processing extensively to obtain better RF performance, it is called *digitally intensive*. Compared to *analog-intensive* architectures, the number of purely analog circuit functions in a digitally intensive mm-wave transceiver is reduced, which results in advantages that conventional digital design flows have over analog design methodologies. Among them are: reduced design cycle times using automated digital implementation tools and flow, ease of testability via built-in self test, on-chip DSP, high-density memory, and automatic functional testing with good fault coverage. Moreover, digitally intensive architectures have lower sensitivity to process/device parameter variability compared to the analog intensive systems. In addition, digital circuits provide reconfigurability to control the operation mode and improve system performance via powerful on-chip calibration techniques, which may reduce silicon area and power dissipation of the SoC.

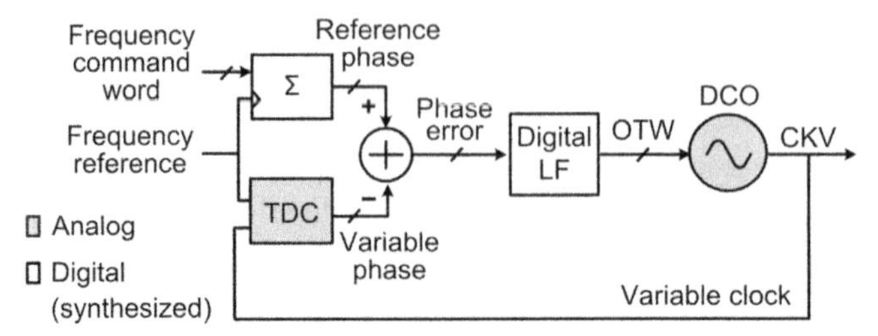

Figure 1.4 Simplified block diagram of an ADPLL.

Note that the "digitally intensive" term doesn't imply that analog/RF design techniques are not important. On the contrary, they are as crucial as before. The overall system performance is usually dominated by a few key analog circuit blocks. The essence of the digitally intensive approach is to make the inputs/outputs (IOs) of the RF/analog building blocks digital so that the system can be modeled and analyzed using the digital design flow, with its many advantages for design throughput and yield. Consequently, it requires RF/analog designers to be conversant with digital circuits and system design, to analyze the system from both analog and digital perspectives, and to collaborate with digital designers.

A good example of a digitally intensive architecture is ADPLL synthesizer shown in Figure 1.4. It contains a digitally controlled oscillator, which may oscillate in the gigahertz range and is controlled by a digital oscillator tuning word (OTW). The gigahertz DCO output (usually followed by an invert-based buffer) has sharp rising/falling edges when implemented in deep-submicron CMOS technologies with f_T above 100 GHz, thus behaving like a digital clock. The time-to-digital converter (TDC) measures and quantizes the time difference between the reference and DCO clock transition edges. Then the digitized phase error is filtered by a digital loop filter (LF) and eventually converted to the OTW in order to tune the oscillator to the desired frequency. Although the DCO and TDC are both analog in nature, all building blocks in the ADPLL are defined as digital at the I/O level, and therefore the loop control circuitry is implemented in a fully digital manner, as illustrated in Figure 1.4.

This ADPLL architecture has been used in mass production for RF connectivity and 2G/3G mobile communications [17]. With the improved RF capability of 65-nm CMOS technology, digitally intensive frequency synthesis could be explored in the mm-wave range, which is over 10× the previously proven frequency range. Such mm-wave

ADPLL increases the reconfigurability of frequency generation in a mm-wave transceiver. Moreover, FM can be incorporated there to form a digital transmitter with the potential for superior modulation quality and lower cost in mass production.

1.2 DESIGN CHALLENGES

To realize low-cost, yet high-performance mm-wave transceivers in CMOS technology, new concepts in IC implementations for ultra-wideband signal generation and mm-wave front-ends are necessary. This book focuses on frequency synthesis, which is critical to many modern communication systems. Due to the high operating frequency, fine frequency resolution, and wideband linear FM requirements, a fully integrated mm-wave frequency synthesizer has various design challenges that are discussed in the following sections.

1.2.1 Toward All-Digital PLL in mm-Wave Regime

Before this very 60-GHz all-digital phase-locked loop (PLL) that will be elaborated in this book, there had been no other reported successful fractional-N synthesizer implementations above 10 GHz for wireless applications that would employ an all-digital approach. There were, however, two reports of digital PLL synthesizers [19,20] operating at 20 and 40 GHz, respectively, used for high-speed serial wireline applications, and numerous successful ADPLL demonstrators operating below 10 GHz for various wireless applications, for example, bluetooth, cellular, WLAN, WiMAX, etc. [21–24]. For low-gigahertz applications, an LC oscillator is normally used to satisfy stringent PN requirements, in which the tuning of the oscillation frequency is achieved via digital control of an array of MOS varactors that operate in flat regions of their $C-V$ curve. It is well-known that the PN of an LC oscillator in the upconverted thermal noise or $1/f^2$ regime (i.e., outside the loop bandwidth of a PLL synthesizer) is inversely proportional to the square of the tank quality factor Q [25]. The tank Q-factor below 10 GHz is dominated by the Q-factor of the inductor, while the varactor Q-factor is normally much higher (e.g., ~100 at 2 GHz) in a 65-nm CMOS technology.

However, this is opposite to the situation at mm-wave frequencies. Figure 1.5 plots the Q-factor and capacitance (C_v) versus bias voltage for n+/n-well and p+/p-well accumulation mode varactors in a 65-nm CMOS technology at minimum gate length [26]. Q-factor varies with

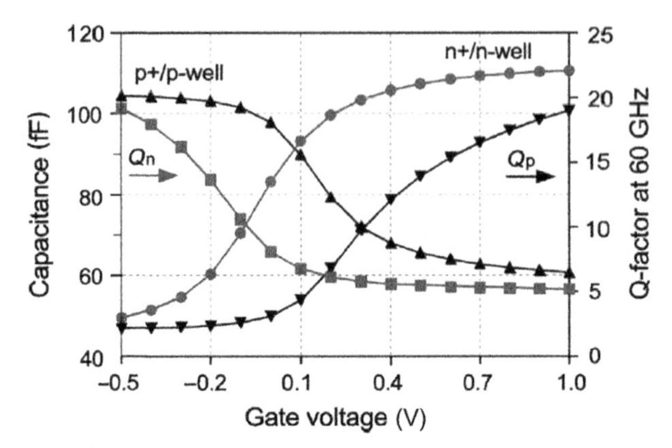

Figure 1.5 Capacitance and Q-factor at 60 GHz versus gate voltage for minimum length, thin-oxide accumulation mode varactors (65-nm CMOS) [26].

bias (due to changing C_v) and is approximately 20 or 5 in the flat regions. While the inductor Q-factor (Q_L) increases with increasing frequency, the Q-factor of capacitors and varactors (Q_C) is inversely proportional to frequency. Therefore, Q of the tank capacitance (varactor plus parasitics) becomes the primary factor limiting the quality of tunable on-chip resonators, and the PN performance of mm-wave oscillators. Power consumption of the oscillator must therefore be increased in order to maintain signal swing and compensate for greater losses in the LC tank.

In addition, a frequency divider chain is necessary to bring the carrier frequency down to a few GHz for further processing by the digital phase detector. There are strong trade-offs between power, chip area (inductors), maximum operating frequency, and operating range in the divider chain design, which normally dissipates more than 50% of the total power in a mm-wave analog PLL [27]. Moreover, the divider's operating range should be aligned with the DCO frequency tuning range in the presence of process, voltage, and temperature (PVT) variations. The divider chain introduces extra delay in the loop and may affect the stability of the PLL. All these bring extra challenges to the design of mm-wave ADPLLs.

Although the digitally intensive nature of the ADPLL permits the fast system-level simulation and verification by an event-driven simulator, the transistor sizing and physical layout of the key "analog-nature" building blocks, such as DCO, divider chain, and TDC have to be "handcrafted" according to the design specifications, and then modeled at the behavioral level for the closed-loop simulations. Compared to design for low-GHz applications, the interconnections between mm-wave frequency building blocks affects the system performance due to parasitic capacitance, losses, and unwanted capacitive and magnetic coupling effects. Therefore, intensive electromagnetic simulation is also required for a successful ADPLL design in the mm-wave regime.

1.2.2 Wide Tuning Range and Fine Frequency Resolution

As mentioned earlier, one major benefit of the 60-GHz band is the 7-GHz worth of unlicensed bandwidth. When the 7-GHz bandwidth is fully employed, the 60-GHz FMCW radar, as shown in Figure 1.3, can theoretically achieve a range resolution as fine as 2 cm. Since PVT variations must be accommodated by the tuning range of the oscillator, a wider than 9-GHz tuning range is desired to ensure full coverage of the entire 60-GHz band. However, a tuning range less than 5% is typically expected for an LC voltage-controlled oscillator (VCO) operating at these frequencies [26].

The tank Q-factor and fractional tuning bandwidth for tanks optimized at each frequency are plotted in Figure 1.6 (schematic shown inset), using simulations in the same 65-nm RF-CMOS technology as Figure 1.5 [26]. In order to construct the tuning range curve, we first

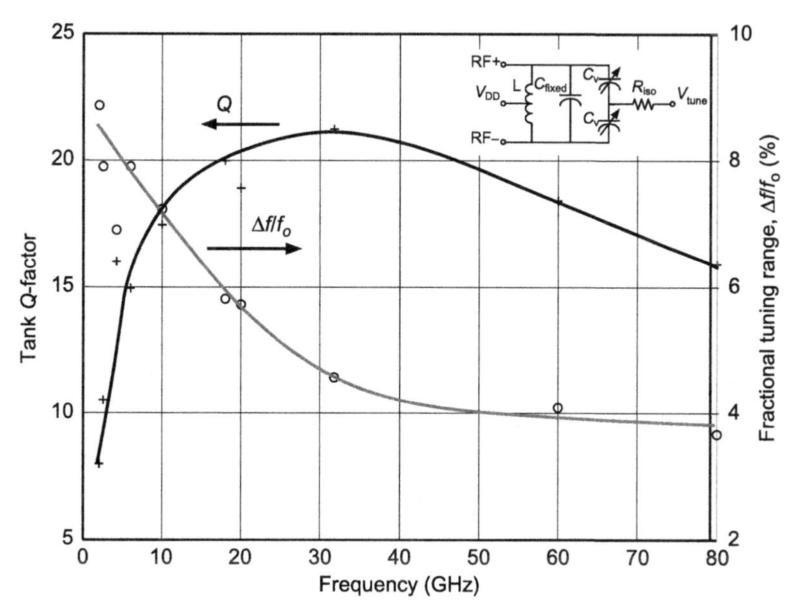

Figure 1.6 Optimum tank Q-factor and fractional tuning range for resonators in 65-nm RF-CMOS from simulation (based on simulations with $C_{fixed} = 20$ fF) [26].

select an inductor that has the highest peak *Q*-factor when driven differentially at each frequency (f_0). The fixed and variable varactor capacitances ($C_{fixed} = 20$ fF and n + /n-well thick oxide varactor with $L = 0.4$ μm) are then added to set the resonance at f_0. The fixed portion of the tank capacitance (C_{fixed}) accounts for wiring interconnects and transistor parasitics. For example, the tunable capacitance ΔC will be 21 fF of the 70.4-fF total tank capacitance (C_0), if a 100-pH tank inductor (L_0) resonant at 60 GHz is assumed. It can be seen that the oscillator tuning range drops to ~5% at higher frequencies. Tuning range may be improved by sacrificing the tank *Q*-factor, i.e., using smaller inductor values and larger varactors. However, the power consumption of the oscillator or the size of core transistor need to be increased for PN performance, which, in the latter case, can introduce more C_{fixed}, thus limiting the achievable tuning range by this tradeoff.

In addition to the aforementioned difficulties in achieving a wide tuning range for a mm-wave LC oscillator, it is also challenging to realize fine frequency tuning under digital control. A mm-wave DCO is the heart of a mm-wave ADPLL. It provides the means to convert digital control words into output frequencies. The lack of a high-resolution DCO has hindered the ADPLL from reaching mm-wave frequencies in the past. The minimum-sized NMOS varactor in a 65-nm CMOS process generates a ΔC of ~40 aF, which results in a frequency resolution of ~17 MHz for a 60-GHz carrier (i.e., $f_0 = \frac{1}{2\pi\sqrt{L_0 C_0}}$, assuming $\frac{\partial f_0}{\partial C_0} \approx \frac{\Delta f_0}{\Delta C_0} = -\frac{f_0}{2C_0}$, thus $\Delta f_0 = \frac{60\ \text{GHz}}{2 \cdot 70\ \text{fF}} \cdot 40\ \text{aF} = 17\ \text{MHz}$). Based on an analysis in Ref. [17], this corresponds to a quantization noise of −62.2 dBc/Hz at 1-MHz offset (reference clock is 40 MHz), which is 28 dB higher than the natural PN of a 60-GHz DCO (e.g., −90 dBc/Hz at 1-MHz offset). Moreover, minimum-sized devices do not track larger devices well, resulting in a mismatch effect inside the tuning bank array. Therefore, new digital fine-tuning techniques need to be developed for mm-wave DCOs to achieve a raw frequency resolution on the order of 1 MHz. ΣΔ dithering of the least significant bits in the DCO tuning bank can be employed to improve the frequency resolution further [17].

1.2.3 Linear Wideband FM

To maximally exploit the available bandwidth (e.g., 7 GHz in the 60-GHz band) allocated at mm-wave frequencies for high data rate communications and high-precision radars, mm-wave transceivers should provide linear wideband FM capability. For example, in a 60-GHz FMCW radar transceiver shown in Figure 1.3, the frequency of the transmit signal is linearly ramping up and down across 5-GHz range. The radar range resolution is determined by Eq. (1.1) and degrades with the sweep nonlinearity [12]. However, in practice, the DCO tuning must be segmented into coarse- and fine-tuning banks, each with different tuning step, K_{DCO} (defined as frequency change per bit) to realize high resolution and a wide tuning range simultaneously. Consequently, the wideband triangular modulation, as shown in Figure 1.3a has to traverse through various tuning banks and rely on linearized K_{DCO} across multiple banks. Moreover, the tuning step mismatches inside each bank also introduce nonlinearities in the FM. Dummy structures can be added in the physical layout to improve the matching performance but they may not be possible at mm-wave frequencies due to the increased parasitics and reduced overall tuning range. Alternatively, digital calibration and compensation techniques should be developed and applied to implement the wideband triangular modulation. In addition, since the Q-factor of the tank also varies from 20 to 5 across the modulation range (Figure 1.5), the output swing of the oscillator may vary by more than 3 dB. The output buffer must compensate for the signal power fluctuation across the modulation range and produce a chirp with a flat output power (e.g., ± dB) at the FMCW transmitter output.

To address the aforementioned concerns and issues for high-performance frequency synthesis at mm-wave frequencies, some new circuits and system architectures arrangements have to be discovered. In this book, alternative design approaches and architecture for mm-wave PLLs are explored. We use a 60-GHz all-digital PLL for FMCW radar application as a design example. The PLL architecture, mm-wave circuit design, and the DSP techniques developed in this particular work can be universally applied to other mm-wave applications that focus on high performance and low cost.

REFERENCES

[1] IEEE P802.11ac/D4.0, Part 11: wireless LAN medium access control and physical layer specifications—amendment 4: enhancements for very high throughput for operation in bands below 6 GHz, November 2012.

[2] P. Adhikari, Understanding millimeter wave wireless communication, 2008 [online]. Available: <http://www.loeacom.com/pdf%20files/L1104-WP_Understanding%20MMWCom.pdf>.

[3] Federal Communications Commission, Title 47: telecommunication part 2—frequency allocations and radio treaty matters; general rules and regulations, May 6, 2008 [online]. Available: <http://www.gpo.gov/fdsys/pkg/CFR-2010-title47-vol1/pdf/CFR-2010-title47-vol1-part2.pdf>.

[4] Federal Communications Commission. Revision of part 15 of the commission's rules regarding ultra-wideband transmission systems, April 22, 2002 [online]. Available: <http://transition.fcc.gov/Bureaus/Engineering_Technology/Orders/2002/fcc02048.pdf>.

[5] International Technology Roadmap for Semiconductors. Radio frequency and analog/mixed-signal technologies for wireless communications, 2009 [online]. Available: <http://www.itrs.net/Links/2009ITRS/2009Chapters_2009Tables/2009_Wireless.pdf>.

[6] H.J. Liebe, D.H. Layton. Millimeter-wave properties of the atmosphere: laboratory studies and propagation modeling, October 1987 [online]. Available: <http://www.its.bldrdoc.gov/pub/ntia-rpt/87-224/index.php>.

[7] IEEE 802.15.3c. Part 15.3: wireless medium access control (MAC) and physical layer (PHY) specifications for high rate wireless personal area networks (WPANs): amendment 2: millimeter-wave based alternative physical layer extension, October 2009.

[8] ECMA International. Standard ECMA-387: high rate 60 GHz PHY, MAC and HDMI PAL, December 2008 [online]. Available: <http://www.ecma-international.org/publications/files/ECMA-ST/ECMA-387.pdf>.

[9] WirelessHD. Overview of wirelessHD specification version 1.0a, August 2009 [online]. Available: <http://www.wirelesshd.org/pdfs/WirelessHD-Specification-Overview-v1%200%204%20Aug09.pdf>.

[10] IEEE 802.11ad. Part 11: wireless LAN medium access control (MAC) and physical layer (PHY) specifications amendment 3: enhancements for very high throughput in the 60 GHz band, 2012.

[11] J.A. Scheer, J.L. Kurtz, Coherent Radar Performance Estimation, Artech House, Inc., Boston, MA, 1993.

[12] G.M Brooker, Understanding millimeter wave FMCW radars, in: Proceedings of International Conference on Sensing Technology, November 2005, pp. 152–157.

[13] Z. Luo, A. Steegen, M. Eller, et al., High performance and low power transistors integrated in 65 nm bulk CMOS technology, in: IEEE International Electron Device Meeting (IEDM) Digest of Technical Papers, 2004, pp. 661–664.

[14] A. Tomkins, R.A. Aroca, T. Yamamoto, S.T. Nicolson, Y. Doi, S.P. Voinigescu, A zero-IF 60 GHz 65 nm CMOS transceiver with direct BPSK modulation demonstrating up to 6 Gb/s data rates over a 2 m wireless link, IEEE J. Solid-State Circuits 44 (8) (2009) 2085–2099.

[15] K. Okada, N. Li, K. Matsushita, K. Bunsen, R. Murakami, A. Musa, et al., A 60-GHz 16QAM/8PSK/QPSK/BPSK direct-conversion transceiver for IEEE802.15.3c, IEEE J. Solid-State Circuits 46 (12) (2011) 2988–3004.

[16] S. Emami, R.F. Wiser, E. Ali, M.G. Forbes, M.Q. Gordon, X. Guan, et al., A 60 GHz CMOS phased-array transceiver pair for multi-Gb/s wireless communications, in: IEEE International Solid-State Circuits Conference Digest of Technical Papers, February 2011, pp. 164–165.

[17] R.B. Staszewski, P.T. Balsara, All-Digital Frequency Synthesizer in Deep-Submicron CMOS, WILEY-Interscience, Hoboken, NJ, 2006.

[18] G.E. Moore, Cramming more components onto integrated circuits, Electron. Mag. (1965) 4. Retrieved November 2006.

[19] A. Rylyakov, J. Tierno, H. Ainspan, J.-O. Plouchart, J. Bulzacchelli, Z.T. Deniz, et al., Bang-bang digital PLLs at 11 and 20 GHz with sub-200fs integrated jitter for

high-speed serial communication applications, in: IEEE International Solid-State Circuits Conference Digest of Technical Papers, February 2009, pp. 94–95,95a.
[20] C.-C. Hung, S.-I. Liu, A 40-GHz fast-locked all-digital phase-locked loop using a modified bang-bang algorithm, IEEE Trans. Circuits Syst. II Express Briefs 58 (6) (2011) 321–325.
[21] R. Staszewski, J. Wallberg, All-digital PLL and transmitter for mobile phones, IEEE J. Solid-State Circuits 40 (12) (2005) 2469–2482.
[22] M. Lee, M.E. Heidari, A.A. Abidi, A low-noise wideband digital phase-locked loop based on a coarse-fine time-to-digital converter with subpicosecond Resolution, IEEE J. Solid-State Circuits 44 (10) (2009) 2808–2816.
[23] L. Vercesi, L. Fanori, F. De Bernardinis, A. Liscidini, R. Castello, A dither-less all digital PLL for cellular transmitters, IEEE J. Solid-State Circuits 47 (8) (2012) 1908–1920.
[24] G. Marzin, S. Levantino, C. Samori, A.L. Lacaita, A 20 Mb/s phase modulator based on a 3.6 GHz digital PLL with −36 dB EVM at 5 mW power, IEEE J. Solid-State Circuits 47 (12) (2012) 2974–2988.
[25] A. Hajimiri, T.H. Lee, A general theory of phase noise in electrical oscillators, IEEE J. Solid-State Circuits 33 (2) (1998) 179–194.
[26] J.R. Long, Y. Zhao, W. Wu, M. Spirito, L. Vera, E. Gordon, Passive circuit technologies for mm-wave wireless systems on silicon, IEEE Trans. Circuits Syst. I Regul. Pap. 59 (8) (2012) 1680–1693.
[27] C. Lee, S.-I. Liu, A 58-to-60.4 GHz frequency synthesizer in 90 nm CMOS, in: IEEE International Solid-State Circuits Conference Digest of Technical Papers, February 2007, pp. 196–197.

CHAPTER 2

Millimeter-Wave Frequency Synthesizers

Contents

2.1 FREQUENCY SYNTHESIZER FUNDAMENTALS

A local oscillator (LO) is required in high-performance radio transceivers irrespective of the architecture. It is employed to translate the RF signal down to an intermediate frequency or baseband in receivers, and vice versa in transmitters. The LO has to be tunable across the RF band and the frequency resolution has to be at least equal to the channel spacing. A frequency synthesizer is typically used as the LO in RF transceivers to overcome the drifts in oscillator frequency due to temperature variations. The synthesizer provides a stable RF carrier with high spectral purity, ideally across a wide frequency span. RF frequency synthesizers remain one of the most challenging blocks in many wireless systems (e.g., mobile communications). The choice of frequency synthesis approach depends on factors such as phase noise (PN), permissible spurious output levels, switching rate, frequency resolution, cost, and complexity.

Millimeter-Wave Digitally Intensive Frequency Generation in CMOS.
DOI: http://dx.doi.org/10.1016/B978-0-12-802207-8.00002-2

2.1.1 PN in Oscillators

An ideal LO operating at angular frequency ω_c, produces a sinusoidal output versus time of the form $y(t) = A \cdot \cos(\omega_c t + \varphi)$, where A is the amplitude and φ is an arbitrary and fixed phase. The zero-crossings occur at integer multiples of the period, $T_c = 2\pi/\omega_c$. In the frequency domain, all of its power is concentrated at a single frequency, ω_c, as shown in Figure 2.1a. However, noise sources inside practical oscillator circuits (e.g., from transistors) perturb the zero crossings randomly. Therefore, both the amplitude and phase vary randomly with time. In most cases, the change in amplitude is removed by a limiting buffer circuit, and therefore only the random deviation of the phase must be considered:

$$y(t) = A_L \cdot \cos(\omega_c t + \varphi_n(t)), \tag{2.1}$$

where $\varphi_n(t)$ is a small, random phase quantity that causes the zero crossings to deviate from integer multiples of T_c. Consequently, the oscillator frequency spectrum spreads around ω_c (Figure 2.1b). The phase function $\varphi_n(t)$ in the time domain is observed as spectral spreading in the frequency domain and is called PN [1].

PN of RF oscillators is normally characterized in the frequency domain. For a small value of the phase fluctuation, $|\varphi_n(t)| \ll 1$ radian, Eq. (2.1) can be simplified to

$$y(t) \approx A \cdot \cos(\omega_c t) - A \cdot \varphi_n(t) \cdot \sin(\omega_c t), \tag{2.2}$$

which means that the spectrum of $\varphi_n(t)$ is frequency-translated to $\pm\omega_c$. Thus, the declining skirts in Figure 2.1b are due to the phase fluctuation $\varphi_n(t)$.

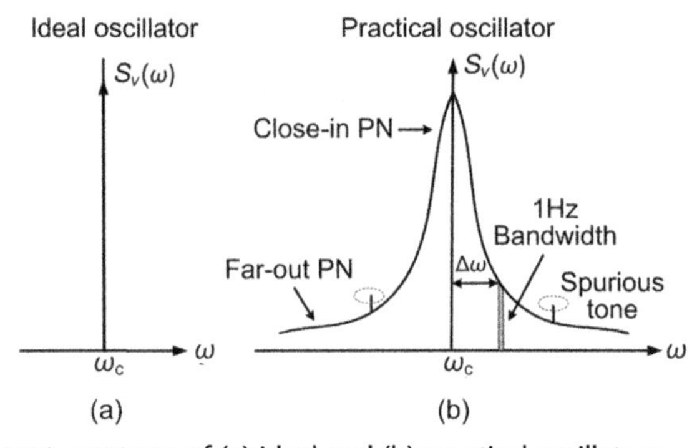

Figure 2.1 Output spectrum of (a) ideal and (b) practical oscillators.

This PN can be quantified by considering a 1-Hz bandwidth at an offset $\Delta\omega$ from the carrier, calculating the noise power, and dividing that result by the carrier power [2].

$$L(\Delta\omega) = 10\log_{10}\frac{\text{noise power in a 1-Hz bandwidth at } \omega_c + \Delta\omega}{\text{carrier power}}. \quad (2.3)$$

This is the single-sided spectral noise density, usually expressed in decibels, with respect to the carrier per hertz bandwidth (i.e., dBc/Hz). The single-sided PN of Eq. (2.3) is one-half of the PN spectrum, which contains both upper- and lower-frequency components ($\omega_c \pm \Delta\omega$). In practice, the PN reaches a constant floor at large frequency offsets (e.g., $\Delta\omega$ larger than a few tens of megahertz in a typical RF oscillator). The region near the carrier is called "close-in" PN and the region far from the carrier is called "far-out" PN, although the border between the two is vague. In this book, "far-out" PN refers to offsets greater than 20 MHz from the carrier.

Figure 2.2 shows a PN spectrum of a typical oscillator. In this log-log plot, the PN in dBc/Hz is plotted against the offset $\Delta\omega$ from the carrier frequency ω_c. The PN profile traverses through $1/\omega^3$, $1/\omega^2$, and $1/\omega^0$ slope regions. The region $1/\omega^2$ is generally referred to as the thermal noise region because it is caused by white or uncorrelated timing fluctuations in the period of oscillation. The $1/f$ flicker noise of electronic devices is also substantial for lower offset frequencies. It gets up-converted and creates the $1/\omega^3$ region. Finally, the $1/\omega^0$ region is typically dominated by the thermal noise added outside the oscillator proper, such as in an output buffer.

In addition, an undesired, systematic fluctuation in the oscillator PN gives rise to a spurious tone. This can be simply explained by considering

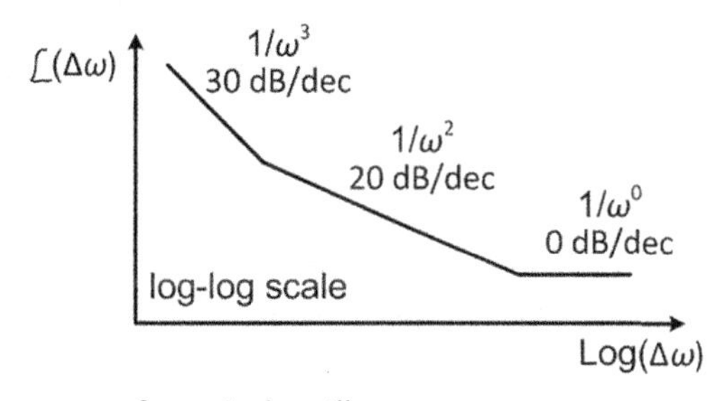

Figure 2.2 PN spectrum of a typical oscillator.

the effect of a single sinusoidal tone on the phase $\varphi_n(t) = \varphi_p \sin(\omega_m t)$. Then, Eq. (2.2) becomes

$$y(t) \approx A \cdot \cos(\omega_c t) + A \cdot \frac{\varphi_p}{2}[\cos(\omega_c + \omega_m)t - \cos(\omega_c - \omega_m)t], \tag{2.4}$$

which indicates that the single sideband PN has a spurious tone at $\Delta\omega = \omega_m$. Spurious tones (or spurs) in a PN spectrum of an oscillator are normally caused by a parasitic coupling from the frequency reference circuitry, phase/frequency detector (PFD) and the charge-pump in classical phase-locked loop (PLL)-based synthesizers. In the time domain, the presence of systematic timing fluctuations in an oscillator waveform represents a periodic timing error. In the frequency domain, it manifests itself as undesired tones in the frequency spectrum. Ideally, the oscillator output spectrum is centered at a single frequency with no spurious tones, as shown in Figure 2.1a. In reality, the presence of spurious tones causes other frequency components to appear in the oscillator output spectrum (see the spurious tones in Figure 2.1b). The amplitude of spurious tones is specified relative to the LO carrier power (i.e., in dBc) at a specific frequency offset from the carrier. It is simply the ratio of the powers of spurious tone and carrier.

2.1.2 Frequency Synthesizer in a Radio Transceiver

The design of RF synthesizers remains one of the most challenging tasks because the synthesizer must operate at a low supply voltage of today's CMOS technologies, while meeting stringent PN and switching transient specifications with low cost and power consumption. We can generally evaluate a synthesizer design by considering the following criteria (in order of importance): PN at a specified offset from the carrier, discrete spurious tones levels, switching speed (defined by the time it takes the synthesizer to hop from its stable state f_0, by a certain frequency excursion Δf, and converge to the new frequency $f_0 \pm \Delta f$), frequency and tuning range, level of integration, and portability between technology nodes.

For wireless applications, a transmitter's PN can cause interference in adjacent bands and distort phase-modulated signals in the process of up/down-conversion, since the PN is indistinguishable from phase or frequency modulation. The resultant constellation points experience small random rotations around the origin, which degrade the error vector magnitude, (EVM; i.e., a measure used to quantify modulation performance of a digital radio transceiver), especially when complex modulation (i.e., I-Q) schemes are employed, for example, 64-QAM [3]. The penalty of PN on a constellation can also be quantified by a poorer bit-error rate in the received signal. In a receiver, PN produced by the LO causes reciprocal mixing of blockers that can severely interfere with the reception of the signal of interest [1]. Consequently, the design budget of the PN of the synthesizer has to be broken down into what the transmitter needs and what the receiver needs. In some cellular standards, the tougher PN requirement rests on the receiver rather than the transmitter due to stringent blocker (i.e., undesired strong signal) requirements. In WLAN IEEE802.11a/g [4], the transmit PN becomes more critical than the receiver's due to the 64-QAM modulation, which has a dense constellation and the bit error rate has to be minimized [1].

For other applications, such as clock data recovery in wireline systems, jitter is commonly used to specify the timing clock performance, where one period of oscillation will differ from the next. The period has an average value T_0 and an instantaneous error ΔT, which indicates the time difference from the average period T_0. The period deviation variance $\sigma_{\Delta T} = \sqrt{\overline{\Delta T^2}}$ is called period jitter. A first-order formula is given in Ref. [5] that relates jitter in the time domain to PN in the frequency domain:

$$L(\Delta\omega) = 10\,\log_{10}\left[\frac{2\pi\omega_c}{\Delta\omega^2}\left(\frac{\sigma_{\Delta T}}{T_0}\right)^2\right], \tag{2.5}$$

in which, $\omega_c = \frac{2\pi}{T_0}$, and $\Delta\omega$ is the offset frequency from the carrier ω_c. Equation (2.5) is applicable to the $1/\omega^2$ up-converted thermal noise, which is the dominant noise mechanism in oscillators.

2.1.3 Methods for Frequency Synthesis

There are three major frequency synthesis techniques: direct analog, direct digital frequency synthesizer (DDFS), and indirect or PLL. The first and second types are routinely found as stand-alone sub-systems. The third type is commonly used in monolithic wireless transceivers, including integer-N and fractional-N operations, and it can be implemented in either an analog-intensive or digitally intensive manner. In an analog-intensive architecture, the phase error generation and oscillator tuning are carried out in the analog domain (e.g., employing a charge-pump and a voltage-controlled oscillator), whereas they are processed in the digital domain in a digitally intensive architecture (e.g., via a time-to-digital converter (TDC), a digital loop filter (LF), and a digitally controlled oscillator (DCO)). A combination of the DDFS and PLL is sometimes employed when a tunable (or modulated) frequency reference is needed at the cost of power consumption. The DDFS is used as a frequency reference of the PLL, replacing the crystal oscillator reference. The operation of DDFS is described briefly in this section, and the PLL is discussed in the next section.

A DDFS system uses logic and memory (look-up table) to construct the desired output signal digitally, and a digital-to-analog converter (DAC) converts it from the digital to analog domains, as shown in Figure 2.3 [6]. The phase accumulator constructs the time-sampled phase at each reference clock cycle and drives a programmable read-only memory, which stores one or more integral number of cycles of a sine wave (or other arbitrary waveform). As the address counter steps through each memory location, the corresponding digital amplitude of the signal at each location drives a DAC, which in turn generates the analog output signal. Therefore, the DDFS method is almost entirely digital, and the amplitude, frequency, and phase are controlled precisely at all times. The digital waveform reconstruction DDFS technique is best suited for implementing wideband transmit modulation as well as fast channel hopping schemes [7].

The spectral purity of the final analog output signal is determined primarily by the DAC and the low-pass filter (LPF), which attenuates spurious frequencies produced by the digital switching. The higher-order harmonics of the DAC output fold back into the Nyquist bandwidth, making them unfilterable, whereas higher-order harmonics at the output of PLL-based synthesizers can be filtered [8]. The PN of a DDFS is determined by the reference clock. The maximum output frequency of a DDFS is limited to around one-third of the reference clock rate. Considering this, and the fact that the phase accumulator operates at the

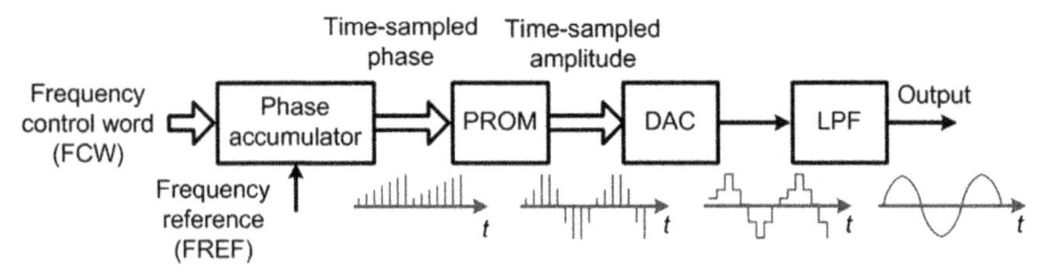

Figure 2.3 Block diagram of a DDFS.

reference clock rate, and DAC and LPF might be difficult to build and power hungry at the gigahertz operational frequency, it is rarely used standalone for RF wireless communications.

2.2 PHASE-LOCKED LOOP

Indirect synthesis using a PLL compares the output phase of a VCO with the phase of a reference signal FREF [9], as shown in Figure 2.4. If the oscillator frequency drifts, the errors are fed back to the VCO tuning input in order to correct the drift using negative feedback. Phase and frequency errors are monitored by the PFD. A LF suppresses spurs synchronized to the reference frequency that are produced by the phase detector so that they do not cause unacceptable frequency modulation in the VCO. In some types of PLLs the phase relationship between FREF and VCO signals (assuming an integer-N operation) is held constant. In other types it is allowed to vary somewhat. However, the frequency is always synchronized (i.e., $f_{VCO} = N \cdot f_{ref}$), otherwise the loop is said to be "out of lock." In general, the division ratio N in the feedback path can be a fractional number. The PLL is named integer-N operation when N is an integer, and otherwise is fractional-N.

A PLL is a feedback control loop and many of the important properties relate to the loop type, which refers to the number of integrators within the loop [10]. Because of the inherent integration in the oscillator, a PLL is always at least type I. Type II PLLs are the most common type in use because the static phase error goes to zero (ideally) when the loop is locked due to a second integration caused by a pole in the LF. Additional non-integrator filtering is often present, contributing additional poles and increasing the order of the loop.

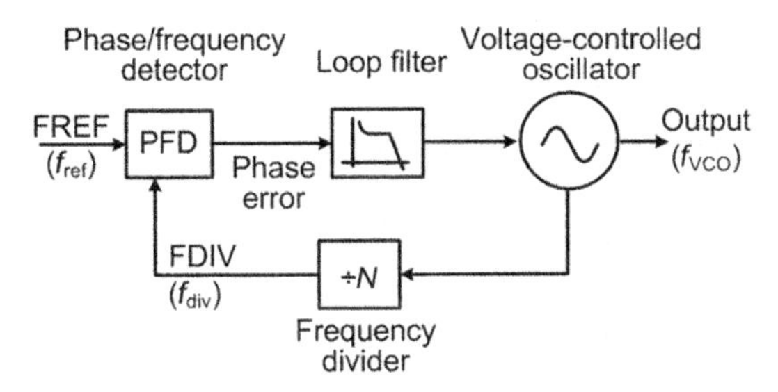

Figure 2.4 Block diagram of a PLL.

2.2.1 Charge-Pump PLL

Figure 2.5 depicts a typical charge-pump PLL. The frequency reference FREF input and the divide-by-N VCO output FDIV are usually squared-up as digital clocks by inverter buffers before being fed to PFD. The PFD estimates the phase difference between the FREF and FDIV by measuring the time difference between the closest rising transitions of these two clocks, and generates either an UP or a DOWN current pulse whose width is proportional to the measured time difference. At the LF, this current pulse is integrated onto capacitors $C_1 + C_2$ to generate a control voltage (V_{Tune}), which sets the average frequency of the VCO. Resistor R_1 allows to provide an instantaneous phase correction without affecting the average frequency. The combination of R_1 and C_2 forms a first-order pole to smoothen the dynamic voltage ripple on V_{Tune} due to the charge-pump and PFD noise, but may make the loop unstable. The combination of R_1 and C_1 forms a zero to stabilize the loop. In a charge-pump PLL, periodic glitches arise from mismatches between the width of UP and DOWN pulses produced by the PFD as well as charge injection and clock feedthrough mismatches between P- and NMOS devices in the charge-pump. These periodic glitches modulate the VCO output frequency, giving rise to frequency spurs.

For wireless applications, fractional-N PLLs are often preferred. The frequency resolution of an integer-N PLL is equal to the reference frequency, which is usually selected to be the same as the channel spacing of an RF transceiver. The PLL bandwidth is normally set at less than 10% of the reference frequency to avoid any significant feedthrough of the reference tone and to maintain loop stability. In response to a change in the frequency command word, the PLL output frequency settles to the programmed value with a time-constant (i.e., settling or switching time)

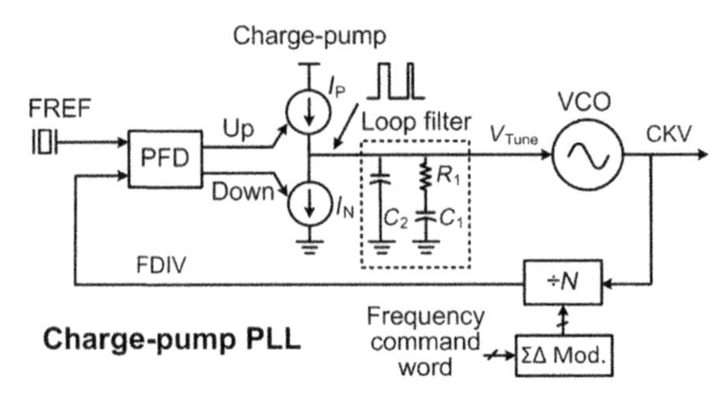

Figure 2.5 Charge-pump-based analog PLL.

inversely proportional to the loop bandwidth. Therefore, narrow loop bandwidths are undesirable because of long switching time, inadequate suppression of VCO PN outside the loop bandwidth, and susceptibility to out-of-band interference from supply and substrate noise.

A fractional-N PLL can achieve an arbitrarily fine time-averaged frequency-division ratio, $N_{\text{ave}} = (N.x)$, by modulating the instantaneous integer division ratio of N and $N+1$, where x corresponds to the fractional part of the frequency-division ratio. The phase detector will operate at a frequency of $f_{\text{ref}} + (0.x/N) \cdot f_{\text{ref}}$, and the phase error of the phase detector causes VCO fractional spurs at multiples of the offset frequency $(0.x) \cdot f_{\text{ref}}$. One widely used method to suppress these spurs is a $\Sigma\Delta$-modulated clock divider (as shown in Figure 2.5) described by Miller and Conley [11] and Riley et al. [12], which trades the reduction in fractional spurs for the increase in the noise floor. In fractional-N synthesizers, the output frequency can increase by fractions of the reference frequency, allowing the latter to be much greater than the required channel spacing. Compared to an integer-N PLL, this allows a wider loop bandwidth at the expense of fractional spurs, resulting in improved loop dynamics and attenuation of the oscillator-induced noise [1]. The loop bandwidth of a fractional-N PLL is normally designed to be a few tens to a few hundreds of kilohertz to suppress the quantization noise of the $\Sigma\Delta$-modulator sufficiently. Quantization noise cancellation techniques were demonstrated in several publications aiming to extend the loop bandwidth without sacrificing the PN performance [13]. In practice, the gain of the DAC in the compensation path is never perfectly matched to that of the signal path through the PFD and charge-pump, so the cancellation of quantization noise is imperfect.

Although the analog charge-pump PLL is still the dominant architecture for RF synthesizers, it is problematic in newer applications, where analog building blocks integrated on a predominantly digital chip pose design and verification challenges. Because of the spur reduction requirements, the analog LF shown in Figure 2.5 usually requires large resistors and capacitors, most likely external to the IC chip, in order to achieve a low PLL bandwidth of several hundreds of kilohertz. Realizing a monolithic capacitance on the order of a few hundred picofarads would require a prohibitively large area if implemented as a metal-oxide-metal capacitor. Implementing it as a MOS capacitor would take less area, but it would likely be unacceptable in scaled CMOS because of its high leakage current, non-linearity, and effect on overall circuit yield in production. The output impedance and the mismatch of the charge-pump currents are not improved with the CMOS scaling. Therefore, the analog-intensive PLL does not lend itself easily to silicon integration and scaling and it lacks portability from one process technology to another.

2.2.2 All-Digital PLL

To strive for improved PLL implementations, going digital appears very attractive and so it has begun to replace the charge-pump PLLs in many mobile applications [14]. Figure 2.6 depicts a simplified block diagram of an all-digital PLL. A digital LF, which is compact and insensitive to a transistor leakage current, replaces the analog LF in Figure 2.5. A DCO replaces the analog tuned VCO, and the phase error is quantized with the aid of a TDC. Thus, many of the aforementioned implementation difficulties associated with charge-pump and analog LF are avoided. Challenges in the design of the DCO and TDC will be discussed in detail in Chapters 4–6.

The ADPLL operates synchronously using fixed-point arithmetic in the phase domain [14] as follows: the reference phase is obtained by accumulating the frequency command word on each FREF rising edge. The integer part of the variable phase is determined by counting the number of rising clock transitions of the DCO oscillator clock CKV, while the fractional part is generated by a TDC which quantizes the time differences between the FREF and DCO clock edges. Then, the sampled variable phase (sampled at FREF rate) is subtracted from the reference phase to obtain digitized phase error samples, which are then filtered by a digital LF and eventually converted to a command word to tune the DCO to the desired frequency and phase. A ΣΔ-modulator is often used to dither the least-significant-bit in the DCO to obtain ultra-fine frequency resolution, for example, 100 Hz of a 60-GHz carrier.

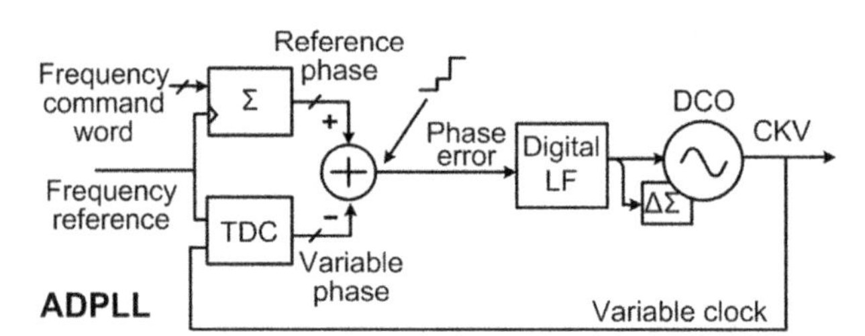

Figure 2.6 TDC-based all-digital PLL.

In an ADPLL, a compact digital LF can be added when necessary to realize a type II loop with higher order to suppress the PN of TDC and reference outside of the PLL closed-loop bandwidth. The digital LF's type and coefficients can be dynamically configured during normal operation without disturbing in phase error (e.g., gear-shifting techniques [15]). The reconfigurability provides a flexible control on PLL closed-loop bandwidth and the loop dynamics. Moreover, the frequency command word in Figure 2.6, which is a fixed-point word to control the PLL frequency, can be augmented dynamically with data to accomplish a frequency or phase modulation of the synthesizer output. The digitally intensive implementation facilitates calibration of the DCO tuning characteristic over PVT variations. Thus, compared to analog PLLs, frequency modulation can be realized with the ADPLL using little additional hardware.

Although the ADPLL has potential advantages over the charge-pump analog PLLs and is amenable to future CMOS processes, a high-performance and low-power implementation remains a challenge. The TDC resolution dominates the PLL PN within the loop bandwidth, while the DCO dominates the PN at higher offset frequencies. In order to obtain an in-band PN of 115 dBc/Hz for a 2.4-GHz carrier, the TDC resolution should be $\sim$1 ps (assume FREF at 26 MHz). To date, only a few TDC-based ADPLLs have achieved this: one uses a two-step TDC with timing amplification [16] and another employs a gated ring oscillator TDC with a first-order noise shaping [17]. Both TDC structures require tremendous design efforts and are sensitive to supply noise. The high-resolution DCO with low PN is another active research area at present.

An alternative to the ADPLL architecture shown in Figure 2.6 is illustrated in Figure 2.7, where a fractional-N divider is used in the feedback path. This architecture mimics the topology of charge-pump PLL

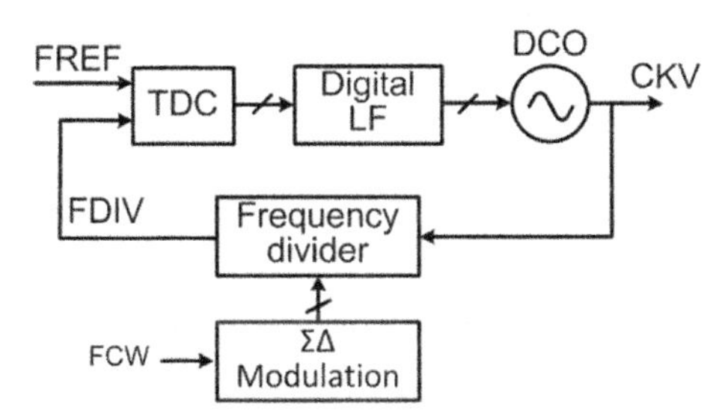

Figure 2.7 All-digital PLL architecture based on frequency divider.

with a $\Sigma\Delta$ multi-modulus divider. Mathematically, the two ADPLL architectures are equivalent and will result in similar performance, but the performance requirements on the individual blocks is different. Figure 2.7 avoids the high-speed counter and register, which are running at RF (e.g., 2 GHz) directly, but it needs the fractional-N divider and a wide-range TDC, which also consume power.

2.3 MILLIMETER-WAVE PLL ARCHITECTURES

The aforementioned frequency synthesis techniques can all be applied to the mm-wave frequency generation. In addition, harmonic generation based on a PLL running at a sub-harmonic of the desired radio frequencies is often considered for mm-wave frequency synthesis due to the challenges of 10× higher output frequency compared to operating at a few gigahertz. Direct generation based on a PLL operating fundamentally at the RF carrier and the harmonic generation approaches are compared in this section with some published design examples. Although these examples are all based on charge-pump analog PLLs, the comparison could also be useful for the digitally intensive architectures because the DCO, dividers, and the multiplier remain analog in nature and are operating at mm-wave frequencies.

The PLL choice for a particular application depends on multiple design aspects: transceiver architecture and frequency plan, frequency tuning range and PN specifications, power consumption, and chip area budget. The signal path and synthesizer are codesigned in a modern transceiver. Synthesizer design has a far-reaching impact on various building blocks in the overall system. Therefore, the aggressive performance targets cannot be met if a synthesizer is designed in isolation, and then interfaced to the rest of the transceiver. Good frequency planning is critical to achieving high system performance at low cost [18], which requires a thorough understanding of both transceiver architectures (e.g., direct conversion, dual conversion) and key building blocks, including the oscillator, divider, multiplier, mixer, filter, etc.

2.3.1 PLL with a Fundamental Oscillator

Figure 2.8 shows a PLL with a fundamental oscillator operating directly at the transceiver channel frequencies (e.g., 60, 77, or 94 GHz) [19–21]. A fundamental oscillator is desirable due to its design simplicity compared to the harmonic generation method. However, the fundamental oscillator

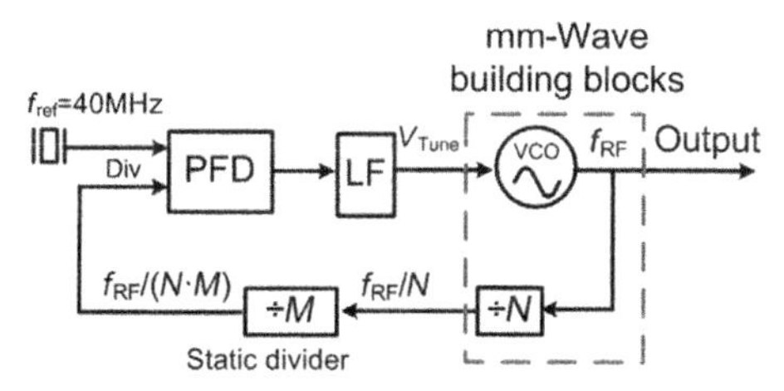

Figure 2.8 mm-Wave PLL with a fundamental oscillator.

(VCO or DCO) should tune across all the channels plus a sufficient margin for PVT variations and provide good PN as well. For example, a 9-GHz range is needed to tune the oscillator across Channels 1–4 of 802.15.3c in the worse case, which is 15% of the center carrier frequency (60 GHz). This is very challenging to achieve according to Figure 1.6. Alternatively, more than one VCO may be utilized to cover the 9-GHz tuning range in the 60-GHz band, for example, in Ref. [19] a high-band VCO operates from 63 GHz to 72 GHz and a low-band VCO from 57 GHz to 64 GHz.

Moreover, the first divider in the prescaler needs to operate across the tuning range of the oscillator. Injection-locked frequency dividers (ILFD; often used in mm-wave bands) may have an insufficient operating range [19]. In addition, misalignment between the VCO tuning and divider operating frequency ranges reduces the usable band of the VCO/divider combination significantly. Therefore, the tuning range of the VCO and the divider must be aligned. One possible calibration method consisting of the VCO sub-band selection and the adjustment of the ILFD locking range was reported in Ref. [22] for a 60-GHz PLL implemented in a 90-nm CMOS.

2.3.2 PLL-Based Harmonic Generation

The harmonic generation topology adopts an oscillator operating at a sub-harmonic of the RF carrier frequency (i.e., $1/N$, N is usually 2 or 3) and then uses either N-push technique to obtain the higher order harmonic or frequency multipliers to reach the desired operating frequency band, as illustrated in Figure 2.9.

The N-push oscillator in Figure 2.9a is a special class of coupled oscillator that explores symmetry to generate the Nth harmonic frequency signal [23]. This is done by coupling oscillators either through a ring-based network or through a star network. When N individual coupled oscillators

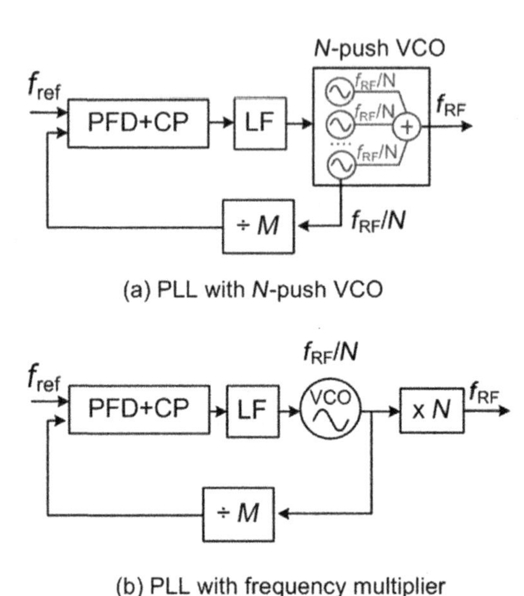

Figure 2.9 PLL-based harmonic generation architectures for mm-wave PLLs.

are synchronized with a progressive phase distribution of $2\pi/N$ radians, the combined output will have all harmonics up to the ($N-1$)th order cancelled and only leave the Nth harmonic. Any phase mismatch between the core oscillators will lead to undesired fundamental feedthrough to the output. When used in a PLL, the N-push oscillator is analogous to a combined oscillator/ILFD, where the phase locking condition is intrinsically satisfied due to the coupling between the oscillators in the N-push operation. An oscillator operating at the lower frequency (i.e., $1/N$ of carrier) would have it easier to achieve wider fractional tuning range and better PN (higher tank Q-factor) compared to a mm-wave fundamental oscillator operating at the carrier frequency (e.g., 94 GHz). Static dividers with wider operating range may be directly used after the oscillator. However, a disadvantage of the N-push oscillator is its single-ended operation, which requires an extra balun to convert to a differential output. Moreover, it is undesired that the harmonic/fundamental signal suppression is sensitive to phase mismatch between the N core oscillators.

Alternatively, Figure 2.9b depicts the mm-wave frequency generation based on a low-frequency PLL followed by a multiplier. The multiplier eases the design requirements on the VCO and divider. When used in a transmitter, LO pulling is minimized as the LO is no longer at the same operating frequency as the output power amplifier. The greater the multiplication factor, the smaller is the chance of LO pulling. The multiplier circuit design detail will be discussed in Chapter 3 and a few frequency multiplier examples selected from the recent literature are listed in Table 3.3.

For PN performance, all harmonic-based systems suffer from PN degradation by 20 $\log_{10}N$, where N is the harmonic order [24]. Nevertheless, the PN of the Nth harmonic output may still be lower than the direct frequency generation, depending on the CMOS process employed and the required carrier frequency (e.g., in the 94-GHz band). The tuning range may also be higher than in the fundamental PLL. However, the power consumption of the harmonic-based synthesizer might be higher than a PLL with the fundamental oscillator (for the same output power) because it is less efficient to generate power at harmonic frequencies. Moreover, rejection of the undesired harmonic/fundamental signal is crucial in the harmonic generation architectures. As discussed above, for the N-push oscillator, this rejection is inherent in the N-push operation but limited by the phase mismatch between the core oscillators. In a multiplier employing a regenerative circuit, for example, the 60-GHz tripler design reported in Ref. [25], the tuned amplifier is responsible for the harmonic rejection, which can be better than a triple-push oscillator at the cost of power consumption [25].

Table 2.1 summarizes the above comparison between direct and harmonic generation methods. As mentioned earlier, the choice of a particular

Table 2.1 Performance comparison between direct and harmonic-based mm-wave frequency synthesizers

	Direct frequency generation	**Harmonic generation with *N*-push oscillator**	**Harmonic generation with multiplier**
Phase noise	Basis	Can be better	Can be better
Tuning range	Basis	Can be better	Basis
Differential output	Basis	Not have	Basis
Frequency pulling	Worse	Basis	Better
Power consumption	Basis	Can be worse	Can be worse
Harmonics rejection	N/A	Worse	Basis

application depends on the transceiver architecture and frequency plan, process technology, frequency tuning range, PN, power consumption, and chip area specifications. For example, a PLL with a fundamental oscillator is preferred for an FMCW radar application, in which a linear frequency modulation is usually obtained via direct PLL modulation [26,27]. Thus, the FMCW chirp generation and transmitting would be simpler and consume less power. The frequency sweep linearity is directly controlled by the feedback loop and not subjected to any potential degradation due to the multiplier or harmonic generation. The LO PN requirement in an FMCW radar is less demanding compared to the communication system since the PN of the transmitted and received signals are correlated, which reduces deleterious effects of the synthesizer PN [28].

From the above analysis of recently published designs in deep-submicron CMOS technology, several guidelines can be derived. A PLL in 65-nm CMOS with a fundamental oscillator is the simplest architecturally and can provide moderate tuning range (~10%) and PN performance (−90 dBc/Hz at 1-MHz offset) with a smallest chip area in the 60-GHz band. For ultra-wide tuning range requirements (>15%) and superior PN (better than −100 dBc/Hz at 1-MHz offset), it is extremely difficult (or even impossible) to meet the goals with a fundamental oscillator at 60-GHz. Hence, the harmonic generation topology has better potential to satisfy the tougher specifications. This doesn't imply that it is easier to design since the multiplier now must have a wide operating band and minimal PN penalty. A few multiplier design examples are discussed in Section 3.3. For a much higher RF carrier, such as 94 GHz or above, the harmonic generation has clearer advantages over the fundamental PLL since the operating frequency is approaching the f_T and $f_{\max}$ of the transistor, which dramatically reduces the design margin. The losses and parasitics in the passive tank are also increasing rapidly and limit the achievable PN and tuning range in a fundamental oscillator above 100 GHz. Another consideration is *IQ* generation, which is needed in an *I*/*Q* transmitter. To generate quadrature signals in the mm-wave regime, a multiplier-based approach would be preferable to a direct quadrature generation at the fundamental frequency due to lower power consumption and *I*/*Q* mismatch [29].

2.4 SUMMARY

This chapter reviews the basics of frequency synthesis as well as mm-wave frequency generation architectures. A TDC-based all-digital

PLL architecture is very promising and tends to be a better PLL, when implemented in scaled CMOS, than the conventional charge-pump PLL. It avoids the bulky analog LF and the problems associated with charge-pump current mismatch, but on the other hand, requires design efforts on the DCO and TDC circuits, which cannot be synthesized via the digital implementation flow. Nevertheless, the digitally intensive nature increases the reconfigurability and testability, and eases the calibration. Two major mm-wave frequency generation approaches, that is, a PLL with a fundamental oscillator and PLL-based harmonic generation are compared with the help of design examples. The key PLL circuit building blocks operating at the mm-wave frequencies will be examined in Chapter 3.

REFERENCES

[1] B. Razavi, RF Microelectronics, second ed., Pearson, United States, 2012.
[2] J. Craninckx, M. Steyaert, Wireless CMOS Frequency Synthesizer Design, Kluwer Academic, Norwell, MA, 1998.
[3] A. Georgiadis, Gain, phase imbalance, and phase noise effects on error vector magnitude, IEEE Trans. Veh. Technol. 53 (2) (2004) 443–449.
[4] IEEE 802.11a/b/g/n. Part 11: Wireless lan medium access control (MAC) and physical layer (PHY) specification, 2012.
[5] T.C. Weigandt, P.R. Gray, Analysis of timing jitter in CMOS ring oscillators, Proc. IEEE Int. Symp. Circuits Syst. 4 (1994) 27–30.
[6] J. Tierney, C. Rader, B. Gold, A digital frequency synthesizer, IEEE Trans. Audio Electroacoust. 19 (1) (1971) 48–57.
[7] L.K. Tan, H. Samueli, A 200-MHz quadrature digital synthesizer/mixer in 0.8-μm CMOS, in: Proceedings of IEEE Custom Integrated Circuits Conference, May 1994, pp. 59–62.
[8] Fundamentals of direct digital synthesis (DDS) [online]. Available: <http://www.analog.com/static/imported-files/tutorials/MT-085.pdf>.
[9] W.F. Egan, Phase Lock Basics, WILEY-Interscience, New York, NY, 1998.
[10] F.M. Gardner, Phaselock Techniques, WILEY-Interscience, Hoboken, NJ, 2004.
[11] B. Miller, B. Conley, A multiple modulator fractional divider, in: 44th Annual Symposium on Frequency Control, May 1990, pp. 559–568.
[12] T.A.D. Riley, M.A. Copeland, T.A. Kwasniewski, Delta-sigma modulation in fractional-*N* frequency synthesis, IEEE J. Solid-State Circuits 28 (5) (1993) 553–559.
[13] A. Swaminathan, K.J. Wang, I. Galton, A wide-bandwidth 2.4 GHz ISM-band fractional-*N* PLL with adaptive phase-noise cancellation, in: ISSCC Digest of Technical Papers, February 2007, pp. 302–303.
[14] R.B. Staszewski, P.T. Balsara, Phase-domain all-digital phase-locked loop, IEEE Trans. Circuits Syst. II Exp. Briefs 52 (3) (2005) 159–163.
[15] R.B. Staszewski, G. Shriki, P.T. Balsara, All-digital PLL with ultrafast acquisition, in: Proceedings of the IEEE Asian Solid-State Circuits Conference, November 2005, pp. 289–292.
[16] M. Lee, M.E. Heidari, A.A. Abidi, A low-noise wideband digital phase-locked loop based on a coarse–fine time-to-digital converter with subpicosecond resolution, IEEE J. Solid-State Circuits 44 (10) (2009) 2808–2816.

[17] M.Z. Straayer, M.H. Perrott, A multi-path gated ring oscillator TDC with first-order noise shaping, IEEE J. Solid-State Circuits 44 (4) (2009) 1089–1098.
[18] K. Okada, et al., A 60 GHz 16QAM/8PSK/QPSK/BPSK direct-conversion transceiver for IEEE 802.15.3c, in: IEEE International Solid-State Circuits Conference Digest of Technical Papers, February 2011, pp. 160–162.
[19] K. Scheir, G. Vandersteen, Y. Rolain, P. Wambacq, A 57-to-66 GHz quadrature PLL in 45 nm digital CMOS, in IEEE Int. Solid-State Circuits Conference Digest of Technical Papers, February 2009, pp. 494–495,495a.
[20] J. Lee, Y.-A. Li, M.-H. Hung, S.-J. Huang, A fully-integrated 77-GHz FMCW radar transceiver in 65-nm CMOS technology, IEEE J. Solid-State Circuits 45 (12) (2010) 2746–2756.
[21] S. Shahramian, A. Hart, A. Tomkins, A.C. Carusone, P. Garcia, P. Chevalier, et al., Design of a dual W- and D-band PLL, IEEE J. Solid-State Circuits 46 (5) (2011) 1011–1022.
[22] T. Shima, K. Miyanaga, K. Takinami, A 60 GHz PLL synthesizer with an injection locked frequency divider using a fast VCO frequency calibration algorithm, in: Proceedings of Asia Pacific Microwave Conference, December 2012, pp. 646–648.
[23] B. Catli, M.M. Hella, Triple-push operation for combined oscillation/divison functionality in millimeter-wave frequency synthesizers, IEEE J. Solid-State Circuits 45 (8) (2010) 1575–1589.
[24] X. Zhang, X. Zhou, A.S. Daryoush, A theoretical and experimental study of the noise behavior of subharmonically injection locked local oscillators, IEEE Trans. Microw Theory Tech 40 (5) (1992) 895–902.
[25] W.L. Chan, J.R. Long, A 56–65 GHz injection-locked frequency tripler with quadrature outputs in 90-nm CMOS, IEEE J. Solid-State Circuits 43 (12) (2008) 2739–2746.
[26] T. Mitomo, N. Ono, H. Hoshino, Y. Yoshihara, O. Watanabe, I. Seto, A 77 GHz 90 nm CMOS transceiver for FMCW radar applications, IEEE J. Solid-State Circuits 45 (4) (2010) 928–937.
[27] C. Wagner, A. Stelzer, H. Jager, PLL architecture for 77-GHz FMCW radar systems with highly-linear ultra-wideband frequency sweeps, in: IEEE International Microwave Symposium Digest, June 2006, pp. 399–402.
[28] G.M. Brooker, Understanding millimetre wave FMCW radars, in: Proceedings of 1st International Conference on Sensing Technology, November 2005, pp. 152–157.
[29] A. Musa, R. Murakami, T. Sato, W. Chaivipas, K. Okada, A. Matsuzawa, A low phase noise quadrature injection locked frequency synthesizer for mm-wave applications, IEEE J. Solid-State Circuits 46 (11) (2011) 2635–2649.

CHAPTER 3

Circuit Design Techniques for mm-Wave Frequency Synthesizer

Contents

3.1 WIDEBAND OSCILLATOR

An LC-based resonator is normally employed to establish the required oscillation frequency (i.e., above 30 GHz) for mm-wave applications. The characteristics of the passive elements define the Q-factor of an LC tank, which directly affects the phase noise of the oscillator. These characteristics are examined in detail in this section. Then, three major frequency tuning techniques that extend the tuning range (TR) of an LC oscillator are compared: switched-capacitor tuning, switched-inductor tuning, and transformer-coupled oscillator. LC-VCOs are used as design examples since only a few mm-wave DCOs have been reported to date. However, most of the oscillator design issues discussed in this section are also applicable to DCOs except for the treatment on analog tuning with varactors. The digital frequency tuning of DCOs and motivations for avoidance of analog varactors at mm-wave will be elaborated in detail in Chapter 5. Distributed oscillators, for example, standing-wave oscillators [1], are not considered here because their large chip area and power consumption mean they are not competitive at the present time [2].

Millimeter-Wave Digitally Intensive Frequency Generation in CMOS.
DOI: http://dx.doi.org/10.1016/B978-0-12-802207-8.00003-4

It is well-known that phase noise of an LC oscillator in the $1/f^2$ regime (i.e., outside the control bandwidth of a PLL synthesizer) is inversely proportional to the square of the tank quality factor [3]. Following the phase noise analysis from Ref. [4] in the $1/f^2$ regime, the phase noise of a cross-coupled CMOS oscillator is given by

$$\mathcal{L}(\Delta\omega) = 10 \cdot \log\left[\frac{(1+\gamma)k_B T}{NA_{\text{tank}}^2 C^2 \Delta\omega^2 R_t}\right]. \quad (3.1)$$

In Eq. (3.1), γ is the transistor's excess noise coefficient ($\sim$2 for thin-oxide devices in 90-nm CMOS), k_B is Boltzmann's constant, T is the absolute temperature (e.g., room temperature = 300 K), A_{tank} is the differential oscillation swing (i.e., 2 V for 1.2-V supply), C is the tank capacitance (typically $\sim$70 fF for 60 GHz resonant frequency), R_t is the parallel equivalent resistance of the LC-tank, N is 2 for differential circuits, and $\Delta\omega$ is the frequency offset of interest (e.g., 1 MHz). Assuming these parameters, the required R_t is 317 Ω for a 60-GHz oscillator to achieve the phase noise of 90 dBc/Hz at 1 MHz offset, which corresponds to a minimum tank Q of 8.4. We assume that at least one of the two core transistors is in saturation in Eq. (3.1), so there will be minimal loading effect on the tank. Noise from the tail current source is also ignored in the above analysis. Considering these factors, it is estimated from Eq. (3.1) that a tank Q-factor better than 10 is required in order to realize an LC-oscillator with −90 dBc/Hz PN at 1-MHz offset from a 60-GHz carrier (i.e., a typical measured phase noise for 60-GHz CMOS LC-VCOs in recent literature). Let's use the minimal tank Q-factor of 10 as our design target and examine the on-chip LC tank's performance.

A spiral inductor and lumped capacitor (partly fixed, partly variable), connected in parallel, are used to implement the LC tank used in most monolithic VCOs. The peak Q-factors from simulations for spiral inductors occupying the copper top metal in a 65-nm RF-CMOS technology [5] (shielded, unshielded, and differential vs. single-ended drive) are plotted in Figure 3.1. The improvement in Q from shielding and using the differential drive versus single-ended is readily apparent from the figure. Note that each data point is associated with a unique inductance value and that the inductance decreases with increasing frequency as noted for a few selected data points. It is also clear that peak Q-factor is proportional to frequency. A single-turn spiral has less inductance per unit length and a lower Q than its transmission line counterpart at mm-wave frequencies, but connection to a differential amplifier is simplified because the inductor terminals lie close to each other in the physical layout.

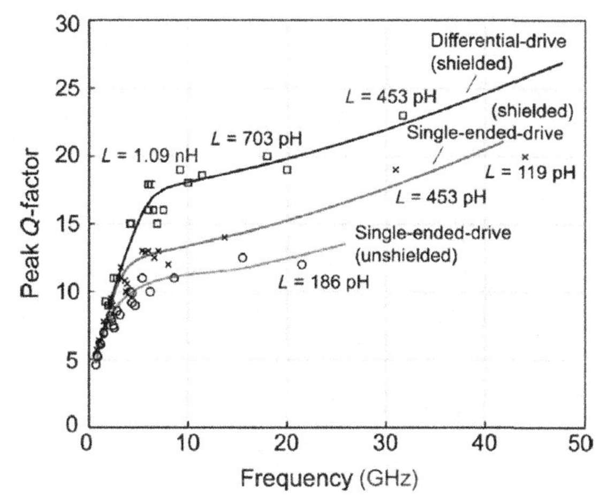

Figure 3.1 Peak Q-factor for monolithic inductors in a 65-nm RF-CMOS technology from simulations [5].

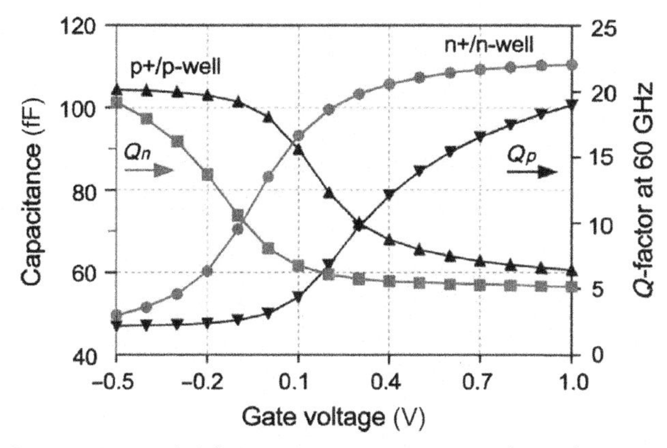

Figure 3.2 Capacitance and Q-factor at 60 GHz vs. gate voltage for minimum length, thin-oxide accumulation mode varactors (65-nm CMOS) [5].

Monolithic capacitors may be either metal-insulator-metal (MIM) or MOS types, with a density of 1–2 fF/mm^2 typical for MIM-type capacitors formed from multiple backend interconnect metals (known as MoM capacitors), or as a two-layer dedicated MIM (usually adding an extra-cost processing option). However, tunable passive VCO resonators require a voltage-variable capacitor (varactor). We have briefly discussed the Q-factor of the n+/n-well and p+/p-well accumulation mode varactors in Chapter 1, as shown in Figure 1.5. It is plotted here again in Figure 3.2. The ratio of maximum to minimum capacitance is approximately 1.5

across the 0—1-V gate bias range imposed by a 1-V supply (e.g., 80 fF at 0 V to 110 fF at 1 V for the n+/n-well varactor). Q-factor varies with bias (due to changing C_v) and frequency as $Q_v = 1/\omega r_v C_v$ (assuming a simple RC model for the varactor). While the inductor Q-factor (Q_L) increases with frequency, the Q-factor of capacitors and varactors (Q_C) is inversely proportional to frequency. The LC resonator quality factor (Q_T) is dominated by the lowest Q component (i.e., $Q_T = Q_C || Q_L$). Therefore, Q_C becomes the factor limiting the quality of tunable on-chip resonators, tunable filters, and the phase noise performance of mm-wave VCOs, as Q_L approaches and exceeds Q_C.

A plot of the time constant ($\tau_v = r_v C_v = 1/\omega Q_C$) for various 100-fF capacitors found in a typical 65-nm RF-CMOS technology is shown in Figure 3.3. Note that the capacitor quality factor Q_C improves with decreasing time constant τ_v. Time constant increases in proportion to increasing the capacitor's series resistance r_v, which models losses in the capacitor dielectric and interconnects. However, scaling of the capacitor to either increase or decrease C_v causes a corresponding change in r_v that keeps τ_v constant (e.g., r_v decreases as C_v increases, and vice versa).

It can be seen that the simulated time constants are frequency-independent, because phenomena such as the skin effect are not included in the simple capacitor model (i.e., in reality r_v is frequency-dependent).

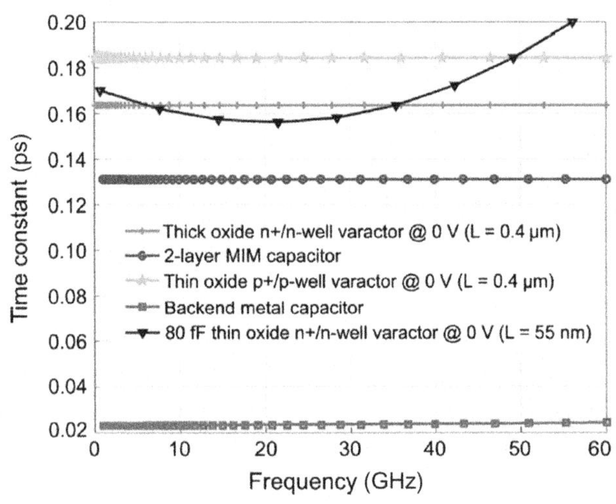

Figure 3.3 Time constant comparisons for monolithic capacitors in 65-nm CMOS. All capacitors are 100 fF except where noted [5].

The one exception is the measured data for an 80-fF thin-oxide varactor plotted in Figure 3.3. The relatively small variation of ±10% in the measured data for τ_v across 60 GHz validates the assumption of a simple RC model for small-valued capacitors. The backend metal capacitor exhibits the lowest τ_v (and hence the best quality factor) for the two fixed MIM capacitor types, which is a consequence of the thick top-metal available in the RF-CMOS technology. The n+/n-well thick-oxide varactor has a lower time constant and hence greater Q-factor across frequency than the p+/p-well varactor when biased at 0 V (gate length $L_g = 0.4\ \mu\text{m}$), as seen from the data plotted in Figure 3.3. This is consistent with the Q-factor data shown for the minimum gate length, thin-oxide accumulation mode varactors in Figure 3.2. Compared to the targeted overall tank Q-factor of 10 derived from Eq. (3.1), there is little design margin, especially when wide TR is required.

A wide oscillator TR is generally required for mm-wave applications. For example, 9-GHz range is needed to tune the oscillator across Channels 1–4 of 802.15.3c to also cover PVT variations. The tuning sensitivity, or change in frequency ($\Delta\omega_o$), of an LC resonator for a change in capacitance ΔC is given by

$$\frac{\Delta\omega_o}{\Delta C} \cong -\frac{\omega_o}{2C_o}, \tag{3.2}$$

where, ω_o is the oscillation frequency determined by $1/\sqrt{LC}$. A 100-pH inductor with a 70.4-fF capacitor resonant at 60 GHz has a very high tuning sensitivity, at −430 MHz/fF according to Eq. (3.2). However, the parasitic capacitance of the inductor, varactor, active devices (e.g., drain-to-bulk and drain-to-gate capacitance from MOS transistors), and interconnect wiring is close to one-half of the total capacitance required for the tank in a typical CMOS VCO (e.g., 20–40 fF fixed capacitance due to parasitics). This limits the variable component of varactor to just a portion of the total tank capacitance C_0, thereby decreasing $\Delta C/C_0$. Hence, the VCO TR ($\Delta\omega$) reduces according to Eq. (3.2). Taking the tank Q-factor requirement (larger than 10 to meet PN target) into account, the oscillator TR will reduce further.

A MOS varactor should be less than 20 fF for a Q-factor better than 10 from simulation in a commercial CMOS 90-nm technology. Assuming the C_{max}/C_{min} of 1.7 for the varactor, the total tunable capacitance ΔC will be 8.2 fF (ΔC = 20 fF-20 fF/1.7). The oscillator TR can be estimated by

$$TR = \frac{f_{\max} - f_{\min}}{f_{\min}} = \frac{1/\sqrt{C_o} - 1/\sqrt{C_o + \Delta C}}{1/\sqrt{C_o + \Delta C}} = \sqrt{1 + \frac{\Delta C}{C_o}} - 1 = 5.7\%, \tag{3.3}$$

which is limited to ~6% in the 60-GHz band.

The restricted TR of varactor-tuned VCOs has sparked research interest in various frequency tuning techniques to extend the operating range of a mm-wave oscillator [6–16].

3.1.1 Oscillator with Switched-Capacitor Tuning

A simplified oscillator schematic with switched-capacitor tuning [6] is plotted in Figure 3.4. The switched-capacitor bank can be implemented as either MOS varactors or MoM capacitors in series with a MOS switch to provide coarse tuning. It can widen the TR compared to oscillators tuned by a single varactor in Ref. [7] (5% TR at 60 GHz) and [8] (1% TR at 77 GHz). The maximum achievable TR is determined by the capacitor tuning ratio (C_{max}/C_{min}), which is normally less than two at 60-GHz when implemented in a 65-nm CMOS. The ratio is limited by the parasitic capacitance from the MOS switch and interconnections. The Q-factor of the capacitor bank becomes a limitation when attempting to

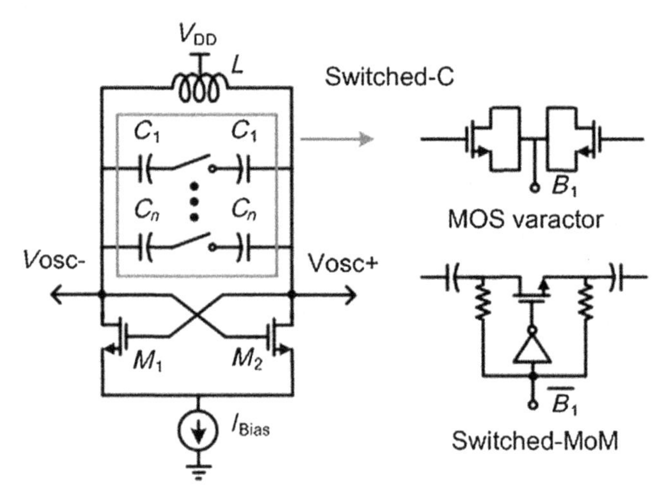

Figure 3.4 Oscillator with switched-capacitor tuning.

meet the phase noise requirement at mm-wave, and thus the TR may be sacrificed to maintain the tank Q-factor. A 60-GHz VCO prototype implemented in 45-nm CMOS achieved 11.6% TR using a 4-bit binary-weighted bank of varactors [9]. In Ref. [10], the switched-capacitor tuning bank was distributed along a transmission line to tune the oscillator from 58 GHz to 64 GHz. The TR of 60-GHz switched-capacitor VCOs reported in the recent literature is consistently less than 15% in CMOS technologies [7—10].

3.1.2 Oscillator with Switched-Inductor Tuning

Switched-inductor tuning can be implemented by switching inductors in series coupled to the primary inductor of the resonator as in Ref. [11] (Figure 3.5), or by switching the mutually coupled inductors to vary the equivalent inductance in the LC-tank as in Ref. [12]. Even though the switched-inductor VCOs described in the literature provide greater TR than switched-capacitor VCOs, they have significantly higher phase noise and/or power dissipation [11,12]. Therefore, magnetic tuning is generally not superior to the basis case of operating a number of separate VCOs except that a lower die area is occupied.

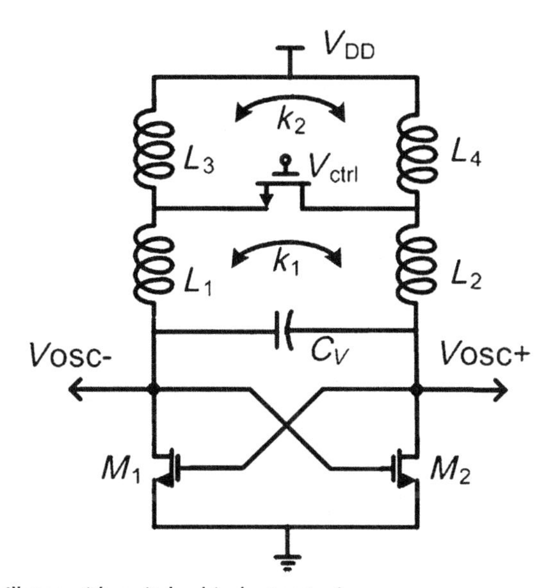

Figure 3.5 Oscillator with switched-inductor tuning.

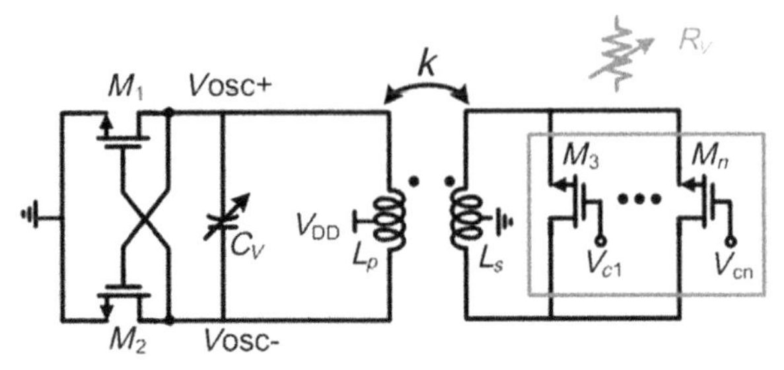

Figure 3.6 Oscillator with transformer tuning via variable load.

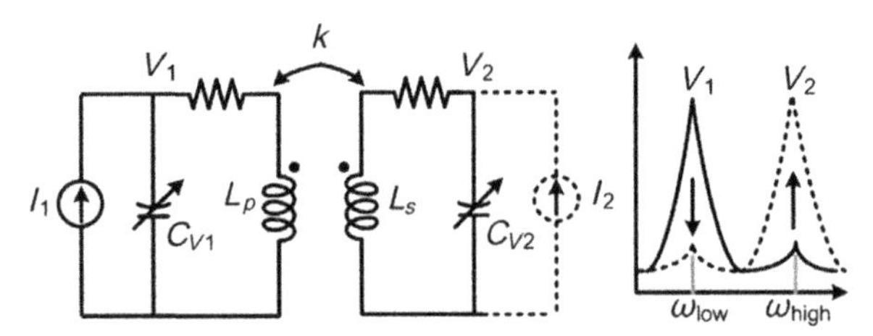

Figure 3.7 Double-tuned double-driven transformer for wideband tuning.

3.1.3 Transformer-Coupled Oscillator

Alternatively, transformer-coupled resonators haven been developed to increase the TR for mm-wave oscillators. Figure 3.6 illustrates frequency tuning by switching the resistive load of the secondary inductor coupled to the primary inductor [13]. In this case, the transformer is excited by driving only one port. The resonator exhibits two resonance points and the VCO can potentially jump between the two oscillating modes. Therefore, it is important to keep the signal level of the desired resonance point much stronger than the other. The VCO tends to oscillate at the resonance frequency with less attenuation because it requires less energy. The transformer can also be double excited, that is, both the primary and secondary ports are excited using current sources I_1 and I_2 as shown in Figure 3.7. The signal amplitudes at the two resonance frequencies can be controlled by changing the amplitude and direction of driving currents I_1 and I_2. Thus, the oscillation frequency is varied by discretely or continuously controlling the current in the secondary winding of a transformer as in Refs. [14–16]. The phase noise of the transformer-coupled oscillator can be better than in the switched-inductor tuning for the same targeted TR as the on-resistance of the switch degrades the Q-factor of the LC tank thus leading to poorer phase noise.

Table 3.1 Performance comparison of various oscillator tuning methods

	Switched-capacitor	Switched-inductor	Transformer-coupled
Tuning range	Basis	+	+
Phase noise	Basis	−	Basis
Area	Basis	−	−

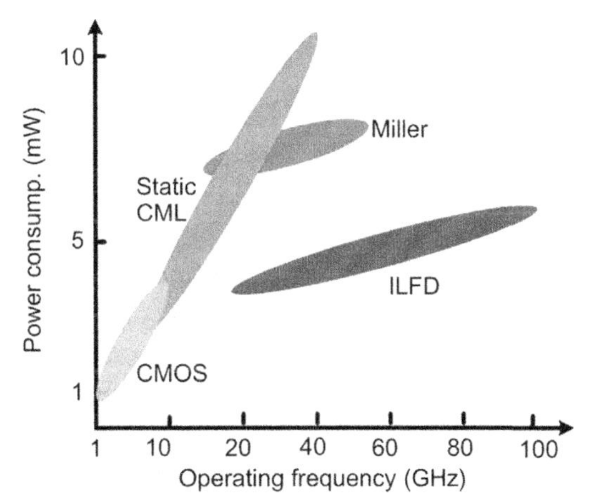

Figure 3.8 Frequency divider topology comparison.

Table 3.1 summarizes the performance of the above three tuning methods. A combination of them is often employed to tune in a continuous fashion across a wide frequency range.

3.2 HIGH-FREQUENCY DIVIDER

Three commonly used divider topologies are reviewed in this section: regenerative (i.e., injection-locked frequency divider, ILFD, and Miller), current-mode logic (CML), and CMOS digital dividers. They are compared, in Figure 3.8, from operating frequency range and power dissipation perspectives. The regenerative dividers (i.e., ILFD and Miller) can operate at very high frequencies (higher than 60 GHz) with moderate power consumption but the operating frequency range is often limited. Static CML dividers have a wide operation range but they usually consume more power when running at 20 GHz or above as compared to ILFD (e.g., more than 10 mA). CMOS digital dividers, which are normally used below 10 GHz consume no static current but require rail-to-rail swing inputs. In a mm-wave frequency synthesizer, a combination of the above divider topologies is normally employed as a trade-off between RF performance, power consumption, and chip area.

3.2.1 Regenerative Frequency Divider

A regenerative frequency divider, also known as a Miller divider, mixes the input signal with a feedback signal from the mixer, as shown in Figure 3.9. The feedback signal there is $f_{in}/2$, which produces the sum and difference frequencies, that is, $f_{in}/2$ and $3f_{in}/2$ at the output of the mixer. A low-pass filter (LPF) removes the higher frequency and the $f_{in}/2$ frequency is amplified and fed back to the mixer. In order to maximize the speed of the Miller divider, it is desirable to employ an LC tank as a band-pass filter to replace the LPF in Figure 3.9.

ILFD is a special kind of regenerative divider. In mm-wave bands, an LC-based ILFD is widely used because of its high-frequency capability and low-power operation [17–21]. An ILFD operates similarly to an injection-locked oscillator, but here the frequency of the input signal is a multiple of the free-running frequency of the oscillator. A basic divide-by-2 structure is shown in Figure 3.10. The self-resonant frequency is approximately one-half of the injected signal frequency, and the locking range is proportional to the injecting current, I_{inj}, and inversely proportional to the oscillator current, I_{osc} [17]. A wider locking range can be achieved either by increasing the size of M_3 or by reducing the capacitive load. Enlarging M_3 increases the power consumption of the ILFD and the circuit that drives M_3. Employing extra inductors (e.g., L_2 in Figure 3.10) to resonate with the parasitic capacitance reduces the capacitive loading at the cost of larger chip area [18]. A division ratio higher than 4 is hardly feasible, since the high-order harmonic signals in a mixing product are inherently low and result in a lower injection efficiency and narrow locking range. An LC-based ILFD consuming a few milliamperes can operate up to 100 GHz [20,21]. The locking range of the ILFD is inversely proportional to the Q-factor of the oscillator tank.

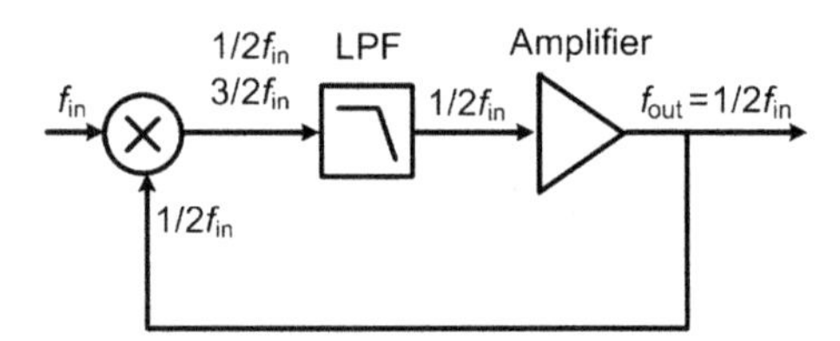

Figure 3.9 Miller frequency divider with LPF.

An alternative research is on inductor-less mm-wave dividers to reduce the chip area occupied by the LC resonant tank. Moreover, an inductor-less design also eases the floor planning when integrated into a PLL as the unwanted magnetic coupling between divider coil and VCO coil is not a concern anymore. An injection-locking divider-by-4 based on an active-load ring oscillator (RO) is illustrated in Figure 3.11 [22]. The prescaler was designed to self-oscillate at around 15 GHz, and to lock at one-fourth of the injected RF frequency. The die area of the ILFD in Ref. [22] is only $80 \times 40\ \mu m^2$, which is less than one spiral inductor of a few hundred picohenry. Although the chip area is dramatically reduced, inductor-less dividers consume more power and normally require extra buffers to be able to drive following stages due to the limited transistor voltage gain at mm-wave frequencies.

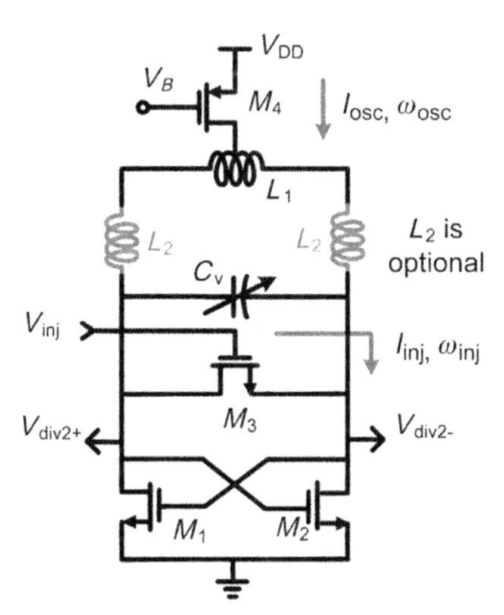

Figure 3.10 Schematic of a divide-by-2 LC-ILFD.

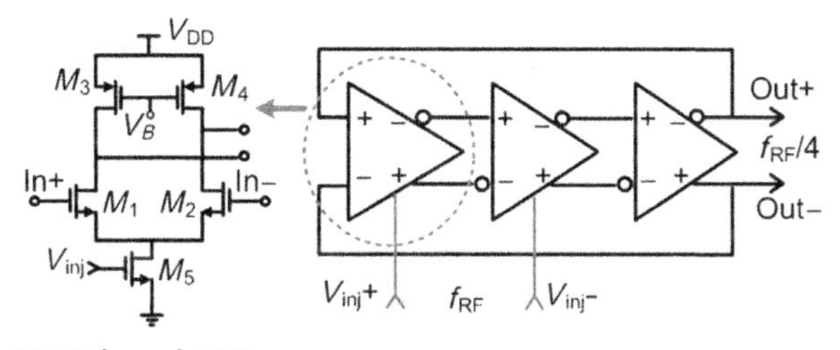

Figure 3.11 RC-RO-based ILFD.

3.2.2 CML Divider

Static CML dividers feature an operating range that is wider than the ILFD, and can extend it down to DC. Figure 3.12 shows the schematic of a typical divide-by-4 CML divider. The static CML divider can be an appropriate choice for input frequencies ranging from 2 GHz to 30 GHz in a 65-nm CMOS technology as a trade-off between area and power consumption. The maximum operating frequency of the static CML divide-by-2 circuit in recent literature is 94 GHz [23] with power consumption of 15.4 mW from a 1.5-V supply. To boost the operating frequency with less power penalty, a dynamic CML latch is often employed at the cost of narrower operating frequency range (a few GHz). Figure 3.13 depicts the schematic of the dynamic CML divider reported in Ref. [24], which achieves division-by-4 up to 70 GHz with 6.5-mW power dissipation in 65-nm CMOS. In order to cover a wide operating frequency

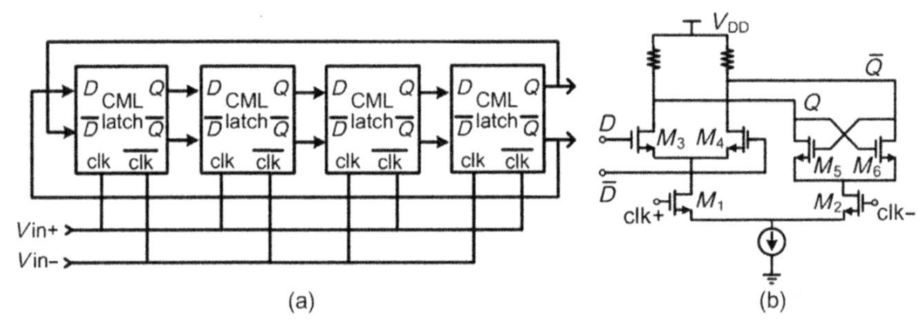

Figure 3.12 (a) Schematic of a divide-by-4 CML prescaler; (b) schematic of the CML latch.

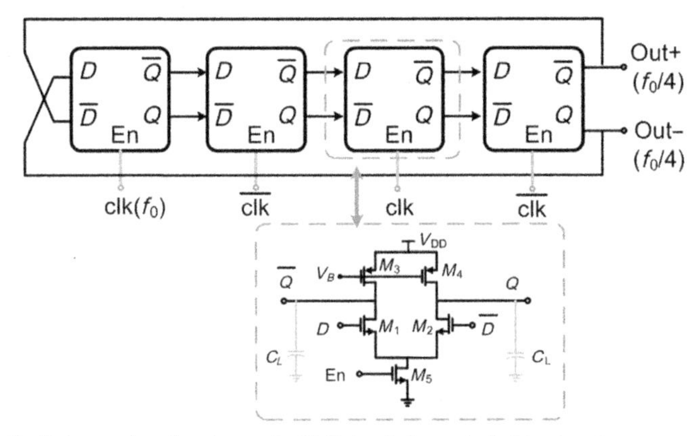

Figure 3.13 Schematic of a dynamic CML latch based divider-by-4.

range from 20 GHz to 70 GHz, the divider is tuned across nine sub-bands each with a locking range of a few gigahertz. Implemented in 32-nm CMOS, a similar dynamic latch-based divide-by-4 obtains 43% locking range for a sub-band centered at 57.5 GHz and consumes only 4.5 mW [25]. Although the dynamic latch-based dividers can operate up to 70 GHz in 65-nm CMOS, they normally provide a higher load on the oscillator or oscillator buffer compared to ILFD when employed in a PLL, which will in turn reduce the frequency TR and increase the overall PLL power consumption.

In summary, the performances of the mm-wave frequency dividers reported in the recent literature are compared in Table 3.2, including the regenerative divider, CML, and dynamic latch-based divider.

3.2.3 Digital CMOS Divider

Although a CML divider is commonly used for low power consumption, quadrature outputs generation, and for handling high input frequency, its broadband phase noise floor is usually limited to around −155 dBc/Hz mainly due to its small internal signal amplitude, long transistor turn-on time, and high conversion gain for both amplitude-modulated and phase-modulated noise present in the circuit. On the other hand, a digital CMOS divider with square-wave internal signals can achieve much lower noise floor and it dissipates no static power. A possible implementation is shown in Figure 3.14 [26]. The only concern is the limited capability of operating at a high input frequency. In the advanced CMOS technology (e.g., 65- and 40-nm CMOS), the CMOS divider can operate up to 10 GHz, which is normally used as the final stage of a mm-wave divider chain.

The circuit in Figure 3.14 requires the complementary clock input (CLK+ and CLK−). Sometimes, it is more desirable to drive the frequency divider single-ended. This type of logic is called true single-phase clocked (TSPC) logic. It is built on the basic digital CMOS divider. It eliminates the differential drive requirement at the expense of more transistors. One such circuit is shown in Figure 3.15 [27]. Transistors M_1 to M_6 form the master latch. Transistors M_7 to M_{12} form the slave latch. M_{13} and M_{14} are simply the inverter to complete the feedback path needed in the toggle latch. When the clock is high, the master latch is transparent while the slave latch is opaque. The opposite is true when the clock is low. The circuit shown in Figure 3.15 is the most basic form. There are more variations and clever designs used to reduce the number of transistors.

Table 3.2 Performance comparison of recently published mm-wave frequency dividers

References	f_{in}/f_{out}	Bandwidth (GHz)	Locking range	Power (mW)	Active area (μm × μm)	Topology	Technology (CMOS) (nm)
[18]	2	48.5–62.9	25.95%	1.65/1.2 V	144 × 112	LC-ILFD	65
[19]	2	59.6–66.96	11.6%	1.6/0.8 V	110 × 150	LC-ILFD	130
[20]	2	128.24–137	6.6%	5.5/1.1 V	320 × 160	LC-ILFD	65
[21]	2	85.2–96.2	12.1%	3.5/1.2 V	140 × 130	LC-ILFD	90
[22]	4	58.4–76	26.2%	25/1.2 V	80 × 40	Inductor-less ILFD	90
[23]	2	64.7[a]–82.4	24%	15.8/1.5 V	40 × 40	Static CML	65 SOI
[24]	4	20–70 (9 sub-bands)	10%–17% (each sub-bands)	6.5/1.0 V	15 × 30	Dynamic latch	65
[25]	4	14–70 (4 sub-bands)	60%–90% (each sub-bands)	4.8/1.0 V	18 × 55	Dynamic latch	32

[a]Lower frequency limited by the W-band test setup.

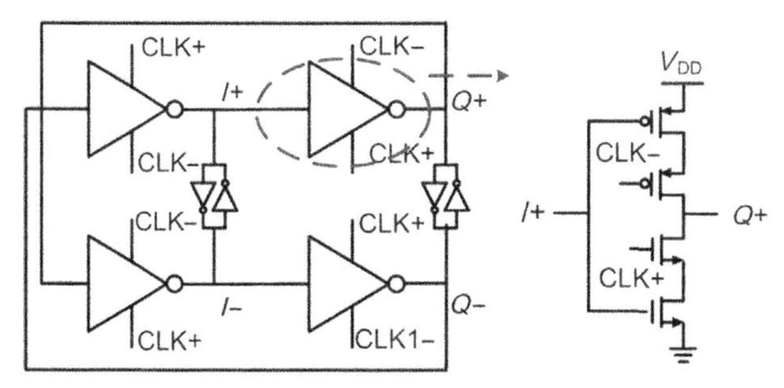

Figure 3.14 Schematic of a digital CMOS divider.

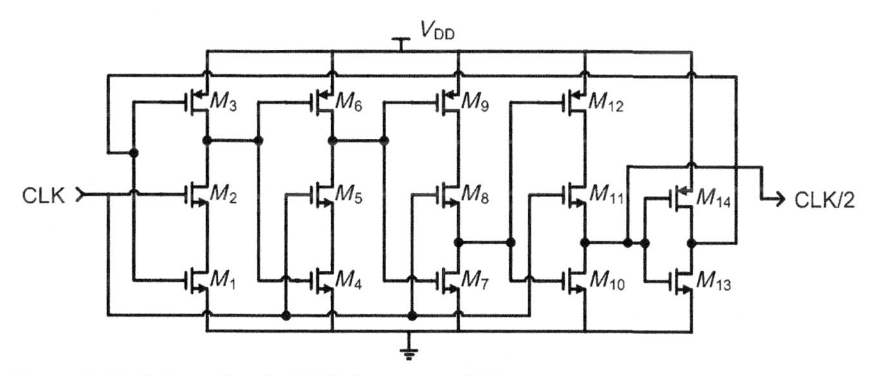

Figure 3.15 Schematic of a TSPC frequency divider.

3.3 FREQUENCY MULTIPLIER

As discussed in Section 2.3, the alternative to fundamental LO generation followed by frequency division in a PLL frequency synthesizer is multiplication. By lowering the oscillation frequency to the range of 15–30 GHz, the VCO would have a larger fractional tuning capability and better phase noise. However, a power-efficient frequency multiplier is required to shift the VCO output up to the mm-wave band (e.g., for a 60-GHz receiver) [28,29].

Frequency multiplier examples selected from the recent literature are listed in Table 3.3. Multipliers can be classified as either active (i.e., with a conversion gain) or passive. The conversion loss of a passive multiplier must be compensated by post-amplification of the multiplied LO output (which consumes power). However, very high output frequencies can be attained (e.g., Refs. [30,31] in Table 3.3). At mm-wave frequencies,

Table 3.3 Performance comparison of recently published mm-wave frequency multipliers

References	f_{in}/f_{out}	Bandwidth (GHz)	Locking range (%)	Power (mW)	Active area (μm × μm)	Topology	Technology (CMOS) (nm)
[28]	2	106–128	18.8	6/1.0 V	120 × 120	LC-IL doubler	65
[29]	3	56–65	14.9	9.6/1.0 V	300 × 300	LC-IL tripler	90
[30]	2	308–328	6.3	Class-B/2.0 V	950 × 340[a]	Differential pair	130
[31]	2	220–275	22.2	40/1.1 V	140 × 160	Traveling-wave double	65
[32]	2	110–125	12.8	$0/V_{bias} = -2$ V	0.21 mm^2	Schottky barrier diode	130

[a]Active area estimated from overall chip size.

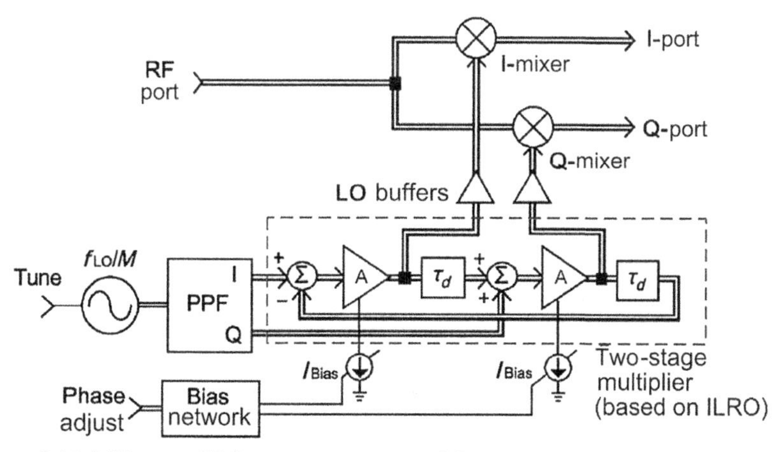

Figure 3.16 MM-wave *I*/*Q* frequency converter [5].

wideband (i.e., low-*Q*) oscillators injection-locked by a passive multiplier are a power-efficient method of multiplication. For example, the injection-locked doubler in Table 3.3 realizes an output between 106 GHz and 128 GHz (19% fractional bandwidth) in a compact chip area with 6-mW power consumption [28].

A block diagram of a power-efficient injection-locked multiplier is shown in Figure 3.16 [5]. The VCO is operating at the desired LO frequency divided by the multiplication factor M (i.e., at f_{LO}/M). A 2-stage RO tuned to free-run at f_{LO} generates quadrature (I and Q) LO at the output of each gain stage (indicated as "A" in Figure 3.16). Locking of the RO to the injected frequency gives multiplication by M. The multiplier is driven by quadrature inputs derived from a polyphase filter to minimize phase error in the $I-Q$ LO outputs. It can be locked by single-phase sub-harmonic injected into only one stage of the ring, but with poorer quadrature output accuracy [33].

The RF signals in Figure. 3.16 are down-converted by the *I*- and *Q*-mixers in a single-sideband mm-wave communication receiver. When configured as an up-converter, the RF port is the single-sideband transmitter output. Suppression of unwanted sidebands in the I/Q frequency converter requires equal amplitude I and Q LOs with low error in their phase quadrature. The phase relationship between I and Q outputs is adjustable (±3°) by varying the bias of each gain stage in the multiplier. LO amplitude variation is suppressed by hard-limiting in the RO and LO buffer stages.

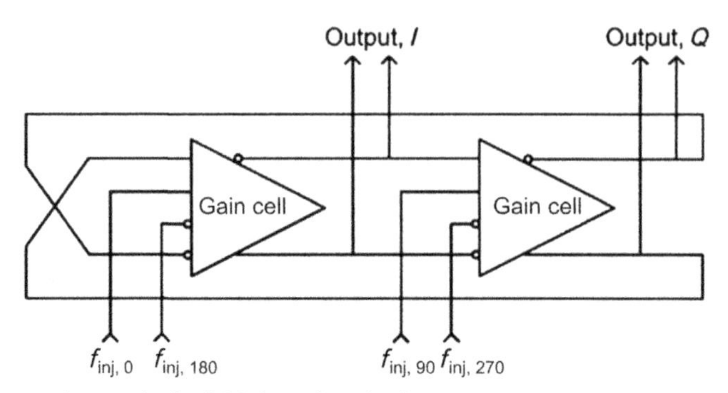

Figure 3.17 Injection-locked RO-based multiplier.

A 60-GHz multiplier-based prototype reported in Ref. [29] employs the above I/Q frequency converter techniques to triple a 20-GHz source to 60-GHz quadrature (i.e., I/Q) outputs. To save power, a RO-based topology was employed, which can be sub-harmonically injection-locked by I/Q inputs ($f_{inj,0}$, $f_{inj,90}$, $f_{inj,180}$, and $f_{inj,270}$) at one-third of the oscillator's free running frequency, as shown in Figure 3.17. The measured injection-locking range is 56.5–64.5 GHz (free-running frequency of 60.6 GHz) when driven by a 320 mV_{p-p} differential source. The measured phase noise penalty between the 20-GHz input and 60-GHz outputs at 100 kHz offset is 9.6 dB, which is close to the ideal for a tripler (i.e., $20\log_{10}[3]$, or 9.54 dB). The circuit consumes 9.6 mW from a 1-V supply, and power consumption could be reduced by at least a factor of 2 (to $\sim$5 mW, or less) if a single-stage LO output instead of I/Q were used.

3.4 SUMMARY

In a mm-wave PLL frequency synthesizer, there are three critical circuit building blocks operating in the mm-wave band; VCO or DCO, frequency divider, and multiplier chain. This chapter examined circuit designs for these critical mm-wave blocks. Contrary to the design trade-off in the low gigahertz range, the tank-Q and VCO (or DCO) phase noise of a mm-wave oscillator are dominated by capacitor/varactor quality, rather than inductor's. The LC tank should be optimized by balancing the frequency dependencies of L and C quality factors. Since process, supply, and temperature variations must also be accommodated by the TR of the oscillator, a VCO with analog tuning becomes marginal with simple varactor tuning. The various capacitive and inductive tuning techniques are reported in the literature to further extend the fractional TR of a VCO to above 10% at 60 GHz. They come at the cost of either degraded phase noise across the band or increased power consumption. Fractional tuning bandwidth improves when a lower operating frequency is chosen, while power consumption and phase noise are optimized by operating close to the peak tank-Q. In addition, the design challenge of the first divider in the PLL feedback loop is also reduced as it operates at lower frequency. However, a power-efficient frequency multiplier is required to shift the VCO output up to the mm-wave band. High-frequency dividers and frequency multiplier circuit examples taken from the recent literature were compared in terms of circuit topology, operating frequency range, power consumption, and chip area.

REFERENCES

[1] F.P. O'Mahony, 10 Gz global clock distribution using coupled standing-wave oscillators, (Ph.D. dissertation), CA August 2003.
[2] J. Chien, S. Member, L. Lu, Design of wide-tuning-range millimeter-wave CMOS VCO with a standing-wave architecture, IEEE J. Solid-State Circuits 42 (9) (2007) 1942–1952.
[3] A. Hajimiri, T.H. Lee, A general theory of phase noise in electrical oscillators, IEEE J. Solid-State Circuits 33 (2) (1998) 179–194.
[4] P. Andreani, X. Wang, L. Vandi, A. Fard, A study of phase noise in colpitts and LC-tank CMOS oscillators, IEEE J. Solid-State Circuits 40 (5) (2005) 1107–1118.
[5] J.R. Long, Y. Zhao, W. Wu, M. Spirito, L. Vera, E. Gordon, Passive circuit technologies for mm-wave wireless systems on silicon, IEEE Trans. Circuits Syst. I, Reg. Papers 59 (8) (2012) 1680–1693.
[6] A. Kral, F. Behbahani, A.A. Abidi, RF-CMOS oscillators with switched tuning, in: Proceedings of IEEE Custom Integrated Circuits Conference, 1998, pp. 555–558.
[7] H. Hoshino, R. Tachibana, T. Mitomo, N. Ono, Y. Yoshihara, A 60-GHz phase-locked loop with inductor-less prescaler in 90-nm CMOS, in: Proceedings of European Solid-State Circuits Conference, vol. 2, September 2007, pp. 472–475.
[8] Y.-A. Li, M.-H. Hung, S.-J. Huang, J. Lee, A fully integrated 77 GHz FMCW radar system in 65 nm CMOS, in: IEEE International Solid-State Circuits Conference Digest of Technical Papers, February 2010, pp. 216–217.
[9] K. Scheir, G. Vandersteen, Y. Rolain, P. Wambacq, A 57-to-66 GHz quadrature PLL in 45 nm digital CMOS, in: IEEE International Solid-State Circuits Conference Digest of Technical Papers, February 2009, pp. 494–495, 495a.
[10] T. LaRocca, J. Liu, F. Wang, D. Murphy, F. Chang, CMOS digital controlled oscillator with embedded DiCAD resonator for 58–64 GHz linear frequency tuning and low phase noise, in: IEEE International Microwave Symposium Digest, June 2009, pp. 685–688.
[11] L. Geynet, E. De Foucauld, P. Vincent, G. Jacquemod, Fully-integrated multi-standard VCOs with switched LC tank and power controlled by body voltage in 130 nm CMOS/SOI, in: Proceedings of IEEE Radio Frequency Integrated Circuits Symposium, June 2006, pp. 1–4.

[12] M. Demirkan, S.P. Bruss, R.R. Spencer, Design of wide tuning-range CMOS VCOS using switched coupled-inductors, IEEE J. Solid-State Circuits 43 (5) (2008) 1156–1163.
[13] T. Lu, C. Yu, W. Chen, C. Wu, Wide tunning range 60 GHz VCO and 40 GHz DCO using single variable inductor, IEEE Trans. Circuits Syst. I, Reg. Papers 60 (2) (2013) 257–267.
[14] K.-C. Kwok, J.R. Long, A 23-to-29 GHz transconductor-tuned VCO MMIC in 0.13 μm CMOS, IEEE J. Solid-State Circuits 42 (12) (2007) 2878–2886.
[15] G. Cusmail, M. Repossi, G. Albasini, F. Svelto, A 3.2-to-7.3 GHz quadrature oscillator with magnetic tuning, in: IEEE International Solid-State Circuits Conference Digest of Technical Papers, pp. 92–93, 2007.
[16] J. Yin, H.C. Luong, A 57.5-to-90.1 GHz magnetically-tuned multi-mode CMOS VCO, Proc. IEEE Custom Integr. Circuits Conf., 2012, pp. 1–4.
[17] M. Tiebout, A CMOS direct injection-locked oscillator topology as high-frequency low-power frequency divider, IEEE J. Solid-State Circuits 39 (7) (2004) 1170–1174.
[18] K. Takatsu, H. Tamura, T. Yamamoto, Y. Doi, K. Kanda, T. Shibasaki, T. Kuroda, A 60-GHz 1.65 mW 25.9% locking range multi-order LC oscillator based injection locked frequency divider in 65 nm CMOS, in: Proceedings of IEEE Custom Integrated Circuits Conference, September 2010, pp. 1–4.
[19] S.J. Rong, A.W.L. Ng, H.C. Luong, 0.9 mW 7 GHz and 1.6 mW 60 GHz frequency dividers with locking-range enhancement in 0.13 μm CMOS, in: IEEE International Solid-State Circuits Conference Digest of Technical Papers, February 2009, pp. 96–97.
[20] B.Y. Lin, K.H. Tsai, S.I. Liu, A 128.24-to-137.00 GHz injection-locked frequency divider in 65 nm CMOS, in: IEEE International Solid-State Circuits Conference Digest of Technical Papers, February 2009, pp. 282–283.
[21] K.H. Tsai, L.C. Cho, J.H. Wu, S.L. Liu, 3.5 mW W-band frequency divider with wide locking range in 90 nm CMOS technology, in: IEEE International Solid-State Circuits Conference Digest of Technical Papers, February 2008, pp. 466–628.
[22] H. Hoshino, R. Tachibana, T. Mitomo, N. Ono, Y. Yoshihara, A 60-GHz phase-locked loop with inductor-less prescaler in 90-nm CMOS, Proc. Eur.Solid-State Circuits Conf 2 (2007) 472–475.
[23] D.D. Kim, J. Kim, C. Cho, A 94 GHz locking hysteresis-assisted and tunable CML static divider in 65 nm SOI CMOS, in: IEEE International Solid-State Circuits Conference Digest of Technical Papers, February 2008, pp. 460–628.
[24] A. Ghilioni, U. Decanis, E. Monaco, A. Mazzanti, F. Svelto, A 6.5 mW inductorless CMOS frequency divider-by-4 operating up to 70 GHz, in: IEEE International Solid-State Circuits Conference Digest of Technical Papers, February 2011, pp. 282–284.
[25] A. Ghilioni, U. Decanis, A. Mazzanti, F. Svelto, A 4.8 mW inductorless CMOS frequency divider-by-4 with more than 60% fractional bandwidth up to 70 GHz, in: Proceedings of IEEE Custom Integrated Circuits Conference (CICC), September 2012, pp. 1,4, 9–12.
[26] C. Hung, R.B. Staszewski, N. Barton, M.-C. Lee, D. Leipold, A digitally controlled oscillator system for SAW-less transmitters in cellular handsets, IEEE J. Solid-State Circuits 41 (5) (2006) 1160–1170.
[27] J.N. Soares Jr., W.A.M. Van Noije, A 1.6 GHz dual modulus prescaler using the extended true-single-phase-clock CMOS circuit technique (E-TSPC), IEEE J. Solid-State Circuits 34 (1) (1999) 97–102.
[28] E. Monaco, M. Pozzoni, F. Svelto, A. Mazzanti, A 6 mW, 115 GHz CMOS injection-locked frequency doubler with differential output, in: Proceedings of IEEE International Conference on IC Design and Technology (ICICDT), June 2010, pp. 236–239.

[29] W.L. Chan, J.R. Long, A 56–65 GHz injection-locked frequency tripler with quadrature outputs in 90-nm CMOS, IEEE J. Solid-State Circuits 43 (12) (2008) 2739–2746.
[30] E. Ojefors, B. Heinemann, U.R. Pfeiffer, Active 220- and 325-GHz frequency multiplier chains in a SiGe HBT technology, IEEE Trans. Microw. Theory Tech 59 (5) (2011) 1311–1318.
[31] O Momeni, E Afshari, Abroadband mm-Wave and terahertz traveling-wave frequency multiplier on CMOS, IEEE J. Solid-State Circuits 46 (12) (2011) 2966–2976.
[32] C. Mao, C.S. Nallani, S. Sankaran, E. Seok, K.O. Kenneth, 125-GHz diode frequency doubler in CMOS, IEEE J. Solid-State Circuits 44 (5) (2009) 1531–1538.
[33] K. Okada, K. Matsushita, K. Bunsen, R. Murakami et al., A 60 GHz 16QAM/8PSK/QPSK/BPSK direct-conversion transceiver for IEEE 802.15.3c, in: IEEE International Solid-State Circuits Conference Digest of Technical Papers, February 2011, pp. 160–162.

CHAPTER 4

All-Digital Phase-Locked Loop

Contents

In Chapter 2, we described briefly the operation of a digitally intensive phase-locked loop (PLL) (also known as all-digital phase-locked loop (ADPLL)), and its potential advantages over charge-pump PLLs. In this chapter, we review the essentials for design of a TDC-based ADPLL. The purpose is to provide the designer with the fundamentals of the ADPLL, such as: how phase correction is performed in the digital domain (Sections 4.1 and 4.2), digital-to-frequency conversion (Section 4.3), the phase/frequency acquisition process with fast locking capability (Section 4.4), ADPLL frequency response (Section 4.5), noise and error sources in the loop (Section 4.6), and behavioral modeling and simulation approaches for the ADPLL (Section 4.7). These basics and the design principles are employed in the mm-wave ADPLL work described in later chapters. For more information on ADPLL operation, refer to *All-Digital Frequency Synthesizer in Deep-Submicron CMOS* [1].

4.1 PHASE-DOMAIN OPERATION

In a PLL, the oscillator (VCO or DCO) output phase, and hence frequency, is corrected periodically by comparison with a stable reference phase as established by the frequency reference (FREF) input of

Millimeter-Wave Digitally Intensive Frequency Generation in CMOS.
DOI: http://dx.doi.org/10.1016/B978-0-12-802207-8.00004-6

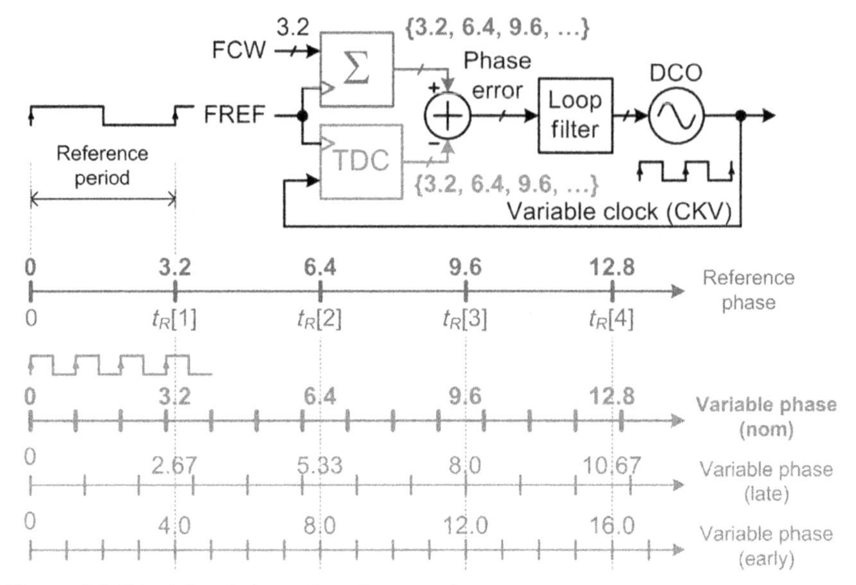

Figure 4.1 Principle of phase-domain operation.

Figure 2.4. In this way, the long-term frequency stability of the synthesizer follows that of the reference (e.g., a crystal oscillator). The major differences between a charge-pump PLL and an all-digital PLL are the phase detection and correction mechanisms, which are performed entirely in the digital domain for the all-digital PLL.

Figure 4.1 depicts the principle of phase-domain operation for an all-digital PLL. The FREF timing is wholly contained in the transition times (i.e., timestamps) of the FREF clock. Of the two possible transition types, only rising clock edges are used here. Likewise, the timing information of the high-frequency variable clock (CKV) is contained in its rising edge timestamps. For the sake of illustration, the frequency command word (FCW) denoting the *expected* frequency multiplicative ratio in this example is 3.2. Since the oscillation period is an inverse of the oscillating frequency, there will be 3.2 clock cycles of CKV per single cycle of FREF.

We first review definitions of the notation used in Figure 4.1. The actual clock period of the RF oscillator (DCO, or VCO in general) output CKV is called T_V, and T_R is the clock period of the FREF. The RF oscillator runs much faster than the available reference clock in a synthesizer, because the multi-GHz RF carrier is orders of magnitude higher in frequency than a typical 10–40-MHz crystal reference. It is assumed that both CKV and FREF behave like digital clocks with sharp positive and negative transition edges. The CKV and FREF clock transition timestamps (i.e., positive transition events, measured in seconds) t_V and t_R are:

$$t_V = iT_V, \tag{4.1}$$

$$t_R = kT_R + t_0, \tag{4.2}$$

respectively. Integers i and k are the CKV and FREF clock transition indices, respectively, and t_0 is an initial time offset (i.e., phase offset) between the two clocks, which is absorbed into the FREF clock. It is convenient in practice to normalize the transition timestamps in terms of actual T_V units, as it is easy to observe and operate on actual CKV clock events. Thus, we can define two dimensionless entities: a variable clock phase and a reference clock phase:

$$\theta_V = \frac{t_V}{T_V}, \tag{4.3}$$

$$\theta_R = \frac{t_R}{T_V}. \tag{4.4}$$

The term θ_V is defined only at CKV-positive transitions and it is indexed by i. Similarly, θ_R is defined only at FREF-positive transitions and is indexed by k. Consequently, we get

$$\theta_V[i] = i, \tag{4.5}$$

$$\theta_R[k] = k\frac{T_R}{T_V} + \frac{t_0}{T_V} = kN + \theta_0. \tag{4.6}$$

Note that both θ_V and θ_R are discrete variables and are dimensionless. Although the CKV and FREF signals are both analog in nature, θ_V and θ_R defined by Eqs. (4.3) and (4.4) are sufficient to represent the real phases of the two clocks for phase detection required in a PLL. Consequently, phase domain operation is based on numerically calculating the phase error between the CKV and FREF $\Phi_E[k] = \theta_R[k] - \theta_V[i]$, with proper alignment of the θ_V and θ_R time samples [2]. The phase error then adjusts the DCO frequency and phase using negative feedback.

A small inconsistency in the previous discussion logic should be noted here. From Eqs. (4.3) and (4.4), the unit of the phase calculation, also called the unit interval, is the CKV clock period T_V. The variable clock (CKV) period, rather than the more stable FREF period is the unit measure of θ_R and θ_V phase quantities, even though the CKV is subject to change due to noise and possible changes in the frequency control word (e.g., due to deliberate frequency modulation as discussed in Chapter 6). Despite this apparent paradox, the system works properly because the error correction mechanism is the difference between these two phase quantities.

4.2 REFERENCE CLOCK RETIMING

It must be recognized that the CKR and CKV clock domains described in Section 4.1 are not entirely synchronous. It is difficult to compare two digital phase values sampled at different time instances t_V and t_R without encountering metastability problems. During frequency acquisition the relationship between these two clock rising edges is not known, and during phase lock the relationship between edges is determined by the fractional FCW. Therefore, it is imperative that phase comparison of the digital word is performed in the same clock domain. This is achieved by oversampling of the FREF clock by the high-rate DCO clock (CKV in Figure 4.2), and using the resulting CKR clock to accumulate the reference phase $\theta_R[k]$ and to sample the high-rate DCO phase $\theta_V[i]$ synchronously. In this way, the digital signal handover between CKR and CKV will be seamless. The CKR clock is thus stripped of the FREF timing information and is used throughout the system. Since the phase comparison is now performed synchronously at the rising edge of CKR, Eqs. (4.5) and (4.6) can be rewritten as

$$\theta_V[k] = \theta_V[i]|_{T_V=[kT_R]} = [kN] = R_V[k], \tag{4.7}$$

$$\theta_R[k] = kN + \theta_0 + \varepsilon[k] = R_R[k] + \varepsilon[k], \tag{4.8}$$

where $\varepsilon[k]$ is the CKV clock edge quantization error in the range of 0 to 1.

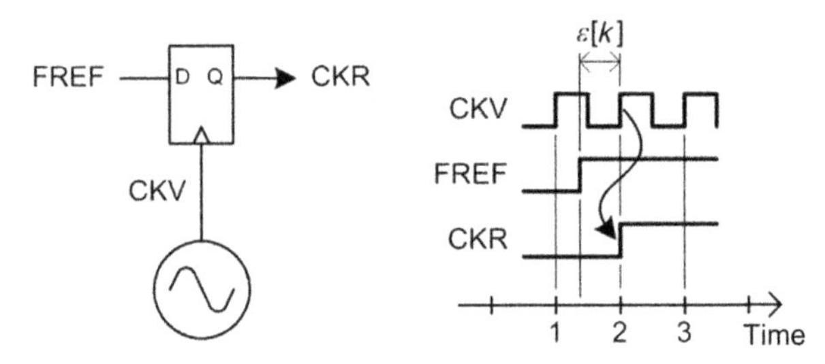

Figure 4.2 Concept of synchronizing clock domains by retiming the FREF.

The $\varepsilon[k]$ could be further estimated and corrected by other means, such as a time-to-digital converter (TDC). This operation is illustrated in Figure 4.2 as an example of integer-domain quantization error $\varepsilon[k]$ as a time difference between the real-valued domain clock edge FREF and its ceiling operation clock edge CKR, which belongs to the integer-valued domain.

Figure 4.3 illustrates conceptual details of the FREF retiming principle. The FREF clock is resampled by the rising and falling edges of the CKV clock to produce QP and QN signals, respectively. These signals are

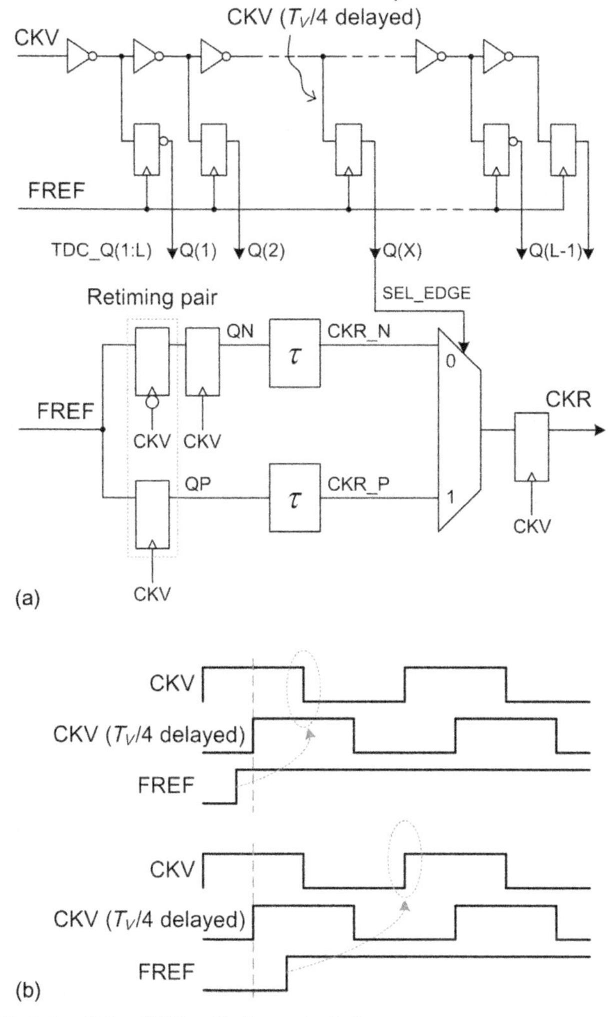

Figure 4.3 Details of the FREF retiming principle.

then delayed by at least a quarter of the DCO clock period (i.e., $\tau > T_V/4$) to produce two CKR candidates: CKR_P and CKR_N. One of these candidates is guaranteed to be free from metastability. The selection is done by the arbiter signal SEL_EDGE, which is obtained by sampling the $T_V/4$ delayed CKV clock by the FREF clock. In practice, the arbiter signal would simply be a programmable TDC tap output Q(X) roughly corresponding to the $T_V/4$ delay. If the rising edge of the delayed CKV clock appears after the rising edge of FREF, then the FREF reclocked by the falling edge of CKV is chosen, as shown at the top timing diagram of Figure 4.3b. The timing diagram at the bottom of Figure 4.3b corresponds to the opposite situation.

The delayed CKV clock could also have its rising edge before the FREF rising edge. In this case, FREF will be retimed by the rising CKV edge, since retiming it on the falling edge of CKV might result in metastability. A special case corresponds to the timing when the arbiter signal is aligned with FREF. In this case, Q(X) might exhibit metastability itself. However, as long as the flip-flop (FF) output Q(X) has a valid logic level, the actual value (logic high or low) does not matter, since both the QP and QN candidates are sufficiently far away ($\sim T_V/4$) from the metastability condition. Certain FFs, such as the current amplifier-biased FF described in Ref. [3], have the property of maintaining a valid output logic level even during the metastability resolution phase.

A careful analysis of the Figure 4.3 circuit metastability behavior is necessary. It involves finding the metastability window and resolution parameters through long SPICE simulations according to the method described in Ref. [4]. The net effect is the final meantime between failures of the system at the worst PVT corner to be on the order of billions of years.

The QP and QN delay described above is needed to maintain the approximate FREF rising edge timing while waiting for the arbiter signal. It can be achieved by retiming the QP and QN signals through FFs.

4.3 DCO GAIN NORMALIZATION AND ESTIMATION

The phase error described above is used to tune the digitally controlled oscillator (DCO), as shown in Figure 4.1. The DCO generates an output with a frequency of oscillation f_V that is determined by the digital oscillator tuning word (OTW) input. In general, f(OTW) is a nonlinear function of the input. However, it could be approximated by a linear transfer function within a limited range of operation. In this case, f(OTW) is a simple gain K_{DCO}. The $f_V = f$(OTW) mapping can be written as

$$f_V = f_0 + \Delta f_V = f_0 + K_{DCO} \cdot \text{OTW}, \tag{4.9}$$

where Δf_V is a deviation from a certain center frequency f_0. f_0 is an adjusted center frequency. The value of Δf_V must be sufficiently small so that the linear approximation is satisfied.

K_{DCO} is specifically defined as a frequency deviation Δf_V (in Hz) from a certain oscillating frequency f_V in response to 1 LSB of the input change. In other words, it is equivalent to the frequency resolution of the fine tuning bank in a DCO. The DCO gain within the linear range of operation can also be expressed as

$$K_{VCO}(f_V) = \frac{\Delta f_V}{\Delta(\text{OTW})}. \tag{4.10}$$

The K_{DCO} should be fairly linear with respect to the input, although it is possible to generalize the DCO gain as also being a function K_{DCO}(OTW) of the specific input:

$$K_{VCO}(f_V, \text{OTW}) = \frac{\Delta f_V}{\Delta(\text{OTW})}. \tag{4.11}$$

Due to its analog nature, the K_{DCO} gain is subject to process and environmental changes that cannot be known precisely. It belongs to one of a few unknown system parameters whose estimate, $\hat{K}_{DCO}$, must be determined. As discussed later, the $\hat{K}_{DCO}$ estimate can be calculated entirely in the digital domain by observing the phase-error responses to the past DCO phase-error corrections. The actual DCO gain estimation involves arithmetic operations, such as multiplication, division, and averaging, and could be performed by dedicated hardware. The oscillator gain dependence on PVT and frequency makes it necessary to estimate it as needed within the actual operating environment.

At a higher level of abstraction, a DCO, together with a DCO gain normalization $f_R/\hat{K}_{DCO}$ multiplier logically comprise a normalized DCO (nDCO), as illustrated in Figure 4.4. The DCO gain normalization decouples the phase and frequency information throughout the system from the process, voltage, and temperature variations that normally affect K_{DCO}. Digital input to the nDCO is a fixed-point normalized tuning word (NTW), whose integer-part LSB corresponds to f_R. The reference

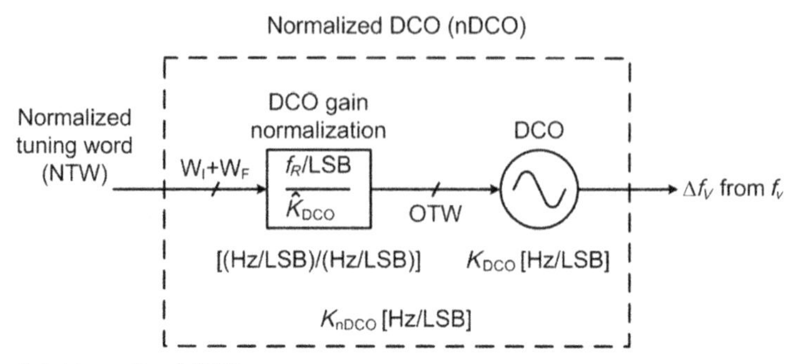

Figure 4.4 Normalized DCO.

frequency is chosen as the normalization factor because it is the master basis for the frequency synthesis. In addition, the clock rate and update operation of this discrete-time system are established by the reference frequency.

The quantity K_{DCO} should be contrasted with the PVT independent oscillator gain K_{nDCO}, which is defined as the frequency deviation (in Hz) of the DCO in response to a 1-LSB change in the integer part of the NTW input. If the DCO gain estimate is exact, $K_{\mathrm{nDCO}} = f_R$; otherwise,

$$K_{\mathrm{nDCO}} = f_R \cdot \frac{K_{\mathrm{DCO}}}{\hat{K}_{\mathrm{DCO}}} = f_R r. \tag{4.12}$$

Dimensionless ratio $r = K_{\mathrm{DCO}}/\hat{K}_{\mathrm{DCO}}$ is a measure of the DCO gain estimation accuracy.

Now let's take a look at the details of DCO gain estimation. According to Eq. (4.10), the DCO gain can be estimated as the ratio of the forced oscillating frequency deviation Δf_V to the steady-state change observed in the OTW, ΔOTW. $\hat{K}_{\mathrm{DCO}}$ is actually used in the denominator of the DCO gain normalization multiplier (Eq. 4.12). This is quite beneficial since the unknown OTW is in the numerator, and the inverse of the forced Δf_V is known and could conveniently be pre-calculated. In this way, the use of dividers is avoided.

Figure 4.5 shows a DCO gain calculation flowchart. After the desired frequency is acquired, M_1 samples of OTW are averaged and the result is stored as OTW_1. Thereafter, a suitable frequency change Δf_V is imposed and the system waits W clock cycles for the PLL to settle. M_2 samples of OTW are then averaged and the result is stored as OTW_2. Finally, the DCO gain estimate $\hat{K}_{\mathrm{DCO}}$ or the normalizing gain $f_R/\hat{K}_{\mathrm{DCO}}$ is computed.

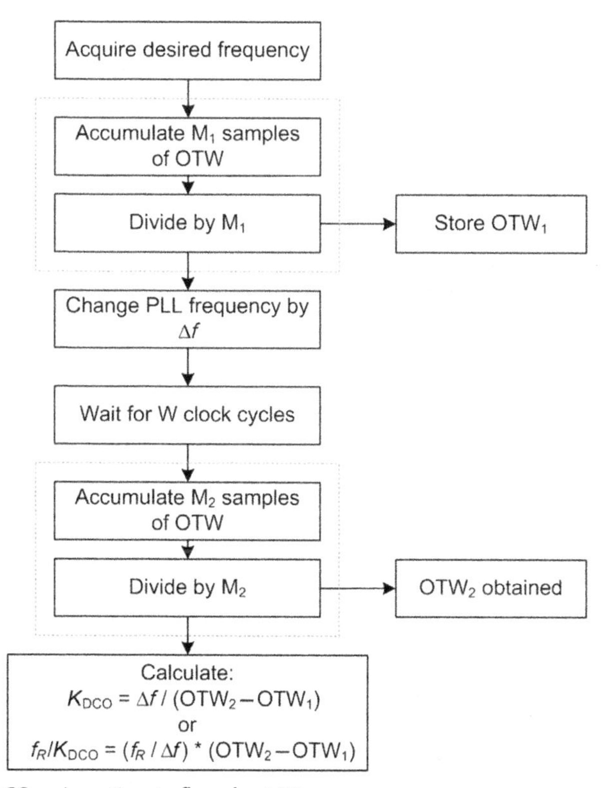

Figure 4.5 DCO gain estimate flowchart [5].

4.4 LOOP GAIN FACTOR AND GEAR SHIFTING OF THE PLL GAIN

4.4.1 Loop Gain Factor

Figure 4.6 assembles the various phase-domain blocks introduced thus far. The variable-phase accumulator $R_V[i]$ counts the number of rising edges of the DCO clock. The variable phase $R_V[i]$ value is sampled by FREF as $R_V[k]$ and adjusted through linear interpolation by means of a TDC system. The reference phase accumulator $R_R[k]$ is obtained by accumulating the FCW at each rising edge of the retimed FREF. The phase detector implements $\Phi_E[k] = \theta_R[k] - \theta_V[k]$ directly. What is still missing at this point is closing of the loop such that the phase error $\Phi_E[k]$ would be used to correct for the frequency and phase drift of the oscillator.

The phase error $\Phi_E[k]$ is instantaneously expressed in units of the reference frequency f_R. The nDCO control is also normalized to f_R. As revealed in Figure 4.6, the frequency transfer function from the nDCO input to the phase detector output is unity, which means that a frequency perturbation Δf_V will be estimated correctly within the quantization resolution $\Delta f_{V,\mathrm{res}}$, in one FREF cycle T_R. For stability reasons, the output of the phase detector cannot be connected directly to the nDCO input because it must be attenuated by a scaling factor α.

Figure 4.7 introduces a scaling factor α that controls attenuation of the phase error before it is applied to the nDCO. In this architecture, the digital loop filter is simply an attenuation factor α, forming a type I PLL. The type I PLL generally features fast locking and its steady-state phase-error signal indicates the frequency offset from the DCO center frequency, which can be expressed as

$$\Delta f_V = -\,\Phi_E \alpha f_R. \tag{4.13}$$

Moreover, the gain factor in Figure 4.7 can be modified to include a second pole at dc, thus giving rise to a type II PLL. This is also known as a proportional-integral (PI) controller. PI control is accomplished in the digital domain by accumulating phase-error samples $\Phi_E[k]$ and scaling them by the integral loop gain ρ (see Figure 4.8). The value of ρ should normally be smaller than α. Contributions from the proportional and integral paths are added together.

A chief advantage of the type II topology is its better filtering capabilities of oscillator noise, leading to improvements in the overall phase-noise performance (see the detailed discussion in Sections 4.5 and 4.6). A type I loop can provide only 20-dB/decade filtering of the DCO phase noise, whereas the type II provides up to 40 dB/decade. As a result, the up-converted DCO flicker noise, which exhibits a 30-dB/decade spectral

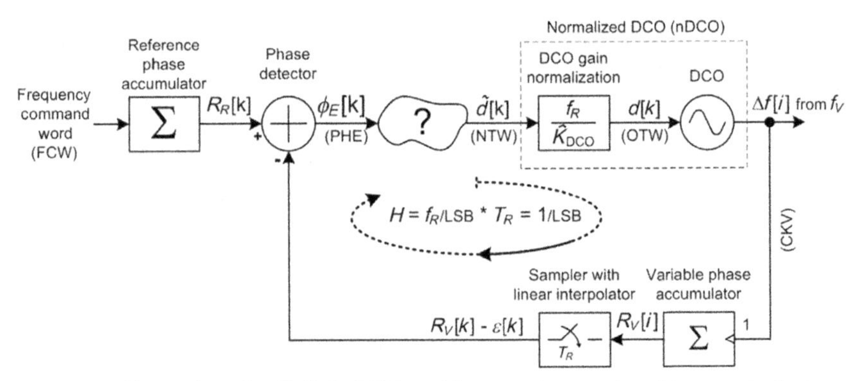

Figure 4.6 Phase-domain all-digital PLL architecture (proportional loop gain factor α remains to be added).

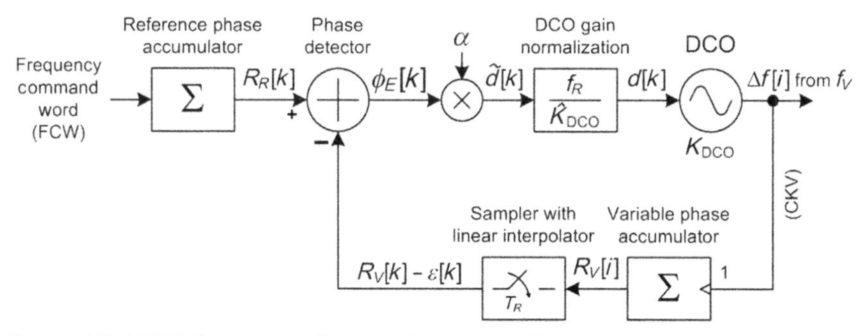

Figure 4.7 ADPLL from a signal-processing perspective.

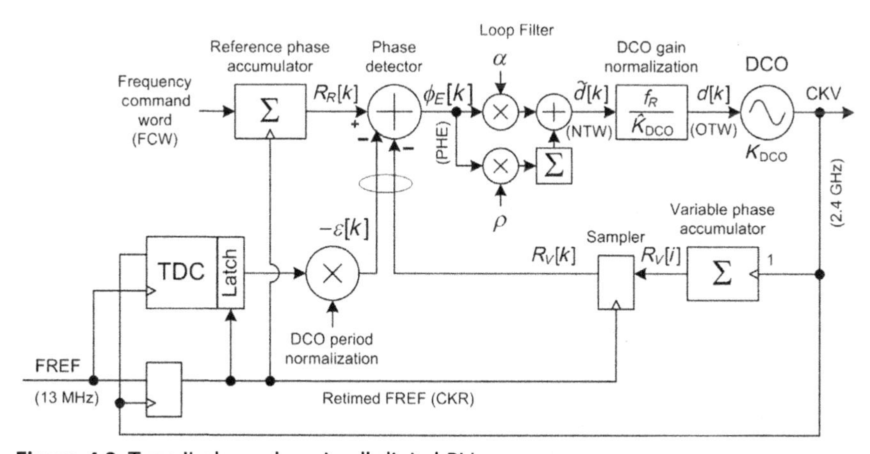

Figure 4.8 Type II phase-domain all-digital PLL.

slope (see Figure 2.2) and could be quite troublesome in nanoscale CMOS, can now be removed completely if the loop bandwidth is large enough. In addition, higher-order loop operation is pontentially advantageous because it provides additional filtering to the phase noise and spurs in the reference path. Another advantage of the type II loop is that it exhibits no steady-state frequency error during ramping of the reference or variable frequencies.

Figure 4.9 shows a loop filter with four single-pole IIR stages and a PI controller. Complex IIR filters could easily become unstable. This problem can be solved by using a cascade of single-pole IIR filters, which are unconditionally stable. The phase delay variation of IIR filters is usually not a problem for fluctuations in the frequencies of interest that are at least one-tenth (or less) of the sampling frequency.

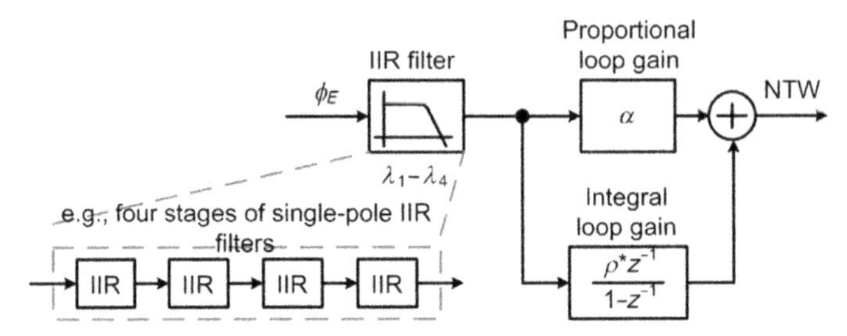

Figure 4.9 Loop filter with single-pole IIR stages and a PI controller.

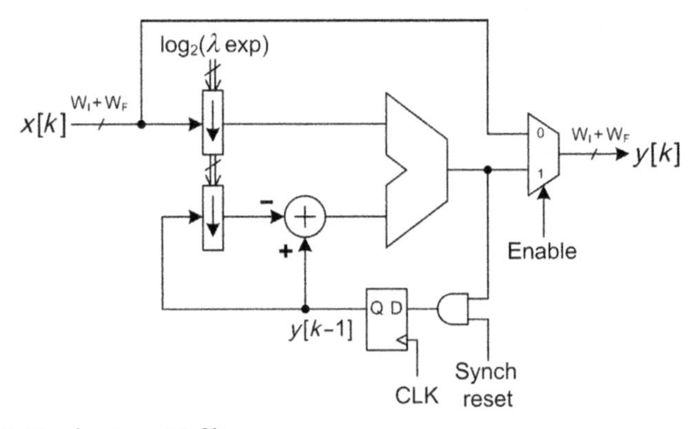

Figure 4.10 Single-stage IIR filter.

Implementation of a single-pole IIR filter is shown in Figure 4.10. The frequency characteristic and, consequently, the pole location are controlled by attenuation factor λ, which is realized as a right-bit-shift operator. The time-domain equation is expressed as

$$y[k] = (1 - \lambda)y[k-1] + \lambda x[k]. \tag{4.14}$$

The z-domain transfer function is expressed as

$$H_{\mathrm{IIR1}}(z) = \frac{\lambda z}{z - (1-\lambda)}, \tag{4.15}$$

and its s-domain representation is

$$H_{\mathrm{IIR1}}(z) = \frac{1 + s/f_R}{1 + s/\lambda f_R}, \tag{4.16}$$

with the 3-dB cutoff frequency

$$f_{\mathrm{BW,IIR1}} = \frac{\lambda}{2\pi} f_R. \tag{4.17}$$

Using analog techniques, a third-order PLL is the highest practical [6,7], mainly for stability and controllability reasons when process and temperature variations are taken into account. However, these restrictions do not exist with digital implementations, and it is possible to create higher-order structures that would provide efficient noise reduction and accurate frequency response [8]. The all-digital loop renders a digital design of the possible loop filter, thus providing benefits in testability, flexibility, and portability to various processes.

4.4.2 Gear Shifting PLL Gain

Two unique intervals with conflicting requirements are readily distinguished during PLL operation. The goal during the first interval is to acquire the desired frequency as soon as possible. The loop bandwidth is made wide at the expense of phase noise and spurs, which are not important during acquisition. In the second interval, which covers the actual Tx and Rx operations, the goal is to maintain or track the desired frequency acquired during the first phase. The loop bandwidth should then be kept lower in order to minimize the effects of phase noise and spurs in the reference path or the PLL output. In addition, higher-order loop operation might be advantageous. These conflicting requirements for the PLL characteristic necessitate a mechanism by which the loop bandwidth is shifted seamlessly once the acquisition phase is completed. The gear-shifting mechanism [9] described here can be used to achieve this.

With gear shifting, activation of the type II loop structure for the purpose of filtering out the DCO flicker noise is deferred until the normal tracking mode. For fast acquisition, the tracking bank varactors are engaged to complete the final frequency settling promptly with a (high) loop gain setting of $\alpha 1$. At the time of switchover into the normal tracking mode, the proportional loop gain α gets reduced to $\alpha 2$ and the integral loop gain is activiated, with its internal accumulator initially set to zero. Unfortunately, the residual dc offset of the preceding mode (in a type I loop the phase error is proportional to the frequency offset) would be an undesired phase error bias which might take a long time to track out. The solution to this problem is to subtract this bias from the phase error before the accumulation operation begins.

Figure 4.11 demonstrates the type II ADPLL loop engagement. After the initial frequency is roughly locked using the PVT-calibration and acquisition modes, the ADPLL activates the tracking bank of the DCO

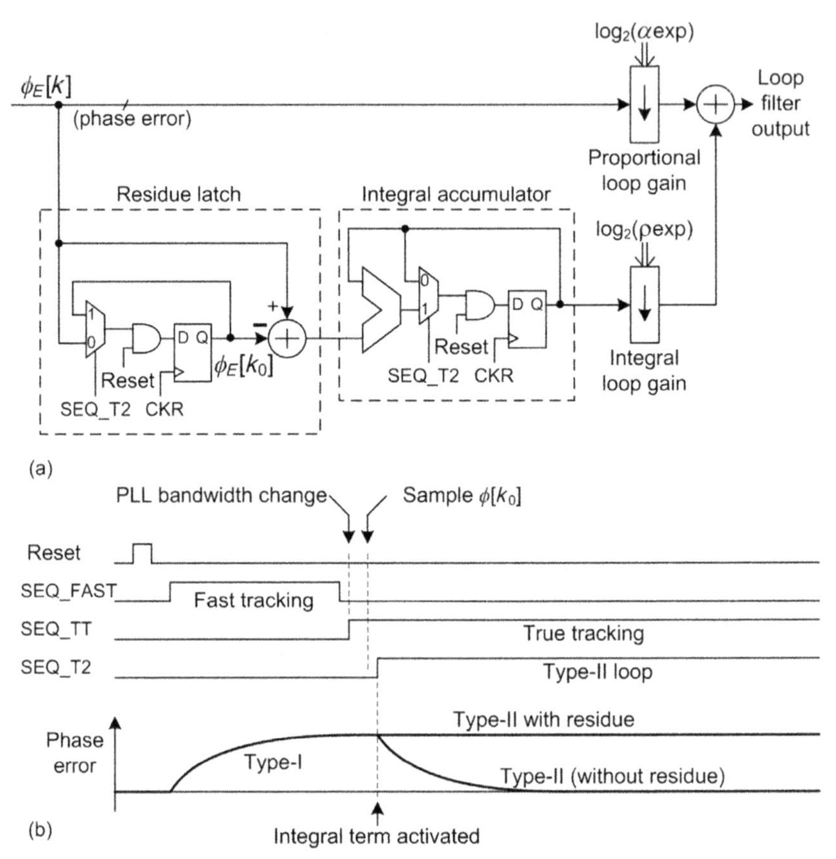

Figure 4.11 Loop filter with dynamically switchable accumulator to perform the type I to type II loop transition: (a) hardware realization; (b) timing of the signals. For clarity, the α gear shifting is not shown.

varactors. At first, the loop bandwidth is quite high ("fast tracking" when SEQ_FAST asserted) and only a proportional loop gain α is used. This allows quick resolution of any frequency quantization error left from the preceding acquisition mode. Then, the normal tracking mode is entered. The following events happen at about the same time: first, the loop bandwidth is narrowed by scaling down the α proportional gain factor (SEQ_TT asserted). This further filters out phase error $\Phi_E[k]$, which then is sampled and stored as $\Phi_E[k_0]$. Finally, the ADPLL type I loop is switched to type II (SEQ_T2 asserted) by activating the integral term ρ. The $\Phi_E[k] - \Phi_E[k_0]$ difference is then accumulated, which gives rise to the type II ADPLL with phase error residue.

Figure 4.11 also reveals the hardware realization of the type II loop filter. All memory elements (registers) are synchronously reset at the beginning of the operation by asserting the synchronous reset control signal. Both α and ρ loop gain factors are implemented in an efficient manner as right bit-shift operations. The residue latch block samples the phase error $\Phi_E[k_0]$ at the beginning of the type II loop operation and outputs the adjusted $\Phi_E[k] - \Phi_E[k_0]$ phase error samples to the integral accumulator block. The conventionally defined type II loop operation ("without residue") could be realized by resetting the synchronous reset signal of the residue latch block, thus forcing $\Phi_E[k_0]$ to zero. The type I loop operation is realized by additionally resetting the integral accumulator.

4.5 PLL FREQUENCY RESPONSE

We have reviewed every key building block in the signal path of the all-digital PLL synthesizer shown in Figure 4.8. The analysis of the PLL frequency response and stability is presented in this section. The closed-loop transfer function derived here will also be used in the next section for noise analysis.

A linearized (s-domain) model for such an all-digital PLL is depicted in Figure 4.12. An extra frequency prescaler ($1/M$ in Figure 4.12) is

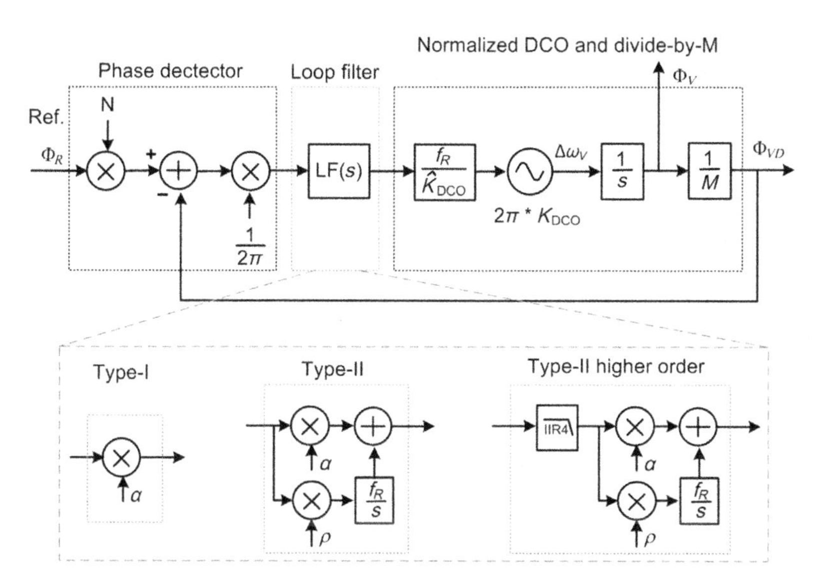

Figure 4.12 Linearized (s-domain) model of an all-digital PLL ($M = 1$ for low-gigahertz application, and $M = 32$ for the 60-GHz PLL example).

usually required in the feedback path of a mm-wave ADPLL to divide down the DCO frequency to a few gigahertz for phase error generation, as described in Section 4.1. Therefore, a $1/M$ stage is added to the linearized model for generality when deriving the PLL frequency response. In a low-gigahertz digital PLL (the focus of this chapter), M is simply set equal to 1.

The LF(s) in Figure 4.12 represents the transfer function of the loop filter in the s-domain. As discussed earlier, it can be type I, with only a proportional gain α (during fast acquisition), or type II with both proportional (α) and integral (ρ) paths, or of higher order, with IIR filters. These LF parameters are programmable and can be dynamically configured during normal PLL operation. The following analysis is focused on type II operation. The stability of the higher-order loop is also examined.

For convenience, we break the loop at Φ_{VD} to calculate the open-loop transfer function $H_{\rm ol}(s)$ for the type II case without IIR filter for simplicity. It is given by

$$\begin{aligned} H_{\rm ol}(s) &= \frac{\Phi_{VD}}{N \cdot \Phi_R} = LF(s) \cdot \frac{f_R}{\hat{K}_{\rm DCO}} \cdot \frac{K_{\rm DCO}}{s} \cdot \frac{1}{M} \\ &= \left(\alpha + \frac{\rho f_R}{s}\right) \cdot \frac{f_R}{\hat{K}_{\rm DCO}} \cdot \frac{K_{\rm DCO}}{s} \cdot \frac{1}{M} \end{aligned} \quad . \tag{4.18}$$

Assuming that the DCO gain is estimated correctly, $H_{\rm ol}(s)$ reduces to

$$H_{\rm ol}(s) = \frac{\Phi_{VD}}{N \cdot \Phi_R} = \left(\alpha + \frac{\rho f_R}{s}\right) \cdot \frac{f_R}{s} \cdot \frac{1}{M}. \tag{4.19}$$

The closed-loop transfer function can then be expressed as

$$H_{\rm cl}(s) = \frac{\Phi_V}{\Phi_R} = \frac{N \cdot M \cdot H_{\rm ol}(s)}{1 + H_{\rm ol}(s)} = N \cdot \frac{\alpha f_R s + \rho f_R^2}{s^2 + \frac{\alpha f_R}{M} s + \frac{\rho}{M} f_R^2}. \tag{4.20}$$

This can be compared to the classical, two-pole system transfer function

$$H(s) = N \cdot \frac{2\zeta\omega_n s + \omega_n^2}{s^2 + 2\zeta\omega_n s + \omega_n^2}, \tag{4.21}$$

where ζ is the damping factor and ω_n is the undamped, natural frequency. The zero lies at $\omega_z = -\omega_n/2\zeta$. Comparing Eq. (4.19) to Eq. (4.20) yields,

$$\omega_n = \sqrt{\frac{\rho}{M}} \cdot f_R, \tag{4.22}$$

and

$$\zeta = \frac{1}{2} \cdot \frac{\alpha}{\sqrt{M\rho}}. \tag{4.23}$$

For a type I loop, the closed-loop transfer function simplifies to

$$H_{\text{cl,type-I}}(s) = N \cdot \frac{\alpha f_R}{s + \frac{\alpha f_R}{M}}, \tag{4.24}$$

and the 3-dB bandwidth of the loop is $f_{BW} = \alpha f_R/(2\pi M)$.

Now, considering the IIR filter in the proportional path of the LF, the transfer function of a one stage IIR filter in s-domain is shown in Eq. (4.16). Each IIR stage has an attenuation factor λ_i ($i = 0,\ldots,3$), and the open-loop transfer function becomes

$$H_{\text{ol,IIR}}(s) = \left(\alpha + \frac{\rho f_R}{s}\right) \cdot \frac{f_R}{s} \cdot \frac{1}{M} \cdot \prod_{i=0}^{3} \frac{1 + \frac{s}{f_R}}{1 + \frac{s}{\lambda f_R}}. \tag{4.25}$$

The closed-loop transfer function can be obtained by substituting for $H_{\text{ol}}(s)$ from Eq. (4.25) into Eq. (4.20).

Adding each stage of IIR filtering introduces a pole at $\omega_{p,i} = j\lambda f_R$ and a zero at $\omega_{z,i} = jf_R$ (from Eq. 4.16), making the loop type-II, but with higher order. This reduces the phase margin in the loop and may cause stability problems. Figure 4.13 shows the magnitude and phase of the open-loop transfer function for the following loop settings:

- $f_V = 62.001$ GHz, $f_R = 100$ MHz, $M = 32$;
- LF: $\alpha = 0.5$, $\rho = 1/2^8$, and $\lambda_i = 0.25, 0.25, 0.25, 0.25$.

The stability analysis results in the following:

1. Open-loop, 0-dB point = 290 kHz; open-loop, $-180°$ point = 2.4 MHz
2. Phase margin = 56.4°; gain margin = 23.9 dB

Being digital, the LF parameters in the ADPLL can be configured easily across a large range in order to provide more flexibility in testing and debugging. The LF can operate type I or type II, and the four stages of the IIR filter may be controlled individually, and switched hitlessly, as described in Section 4.4.2.

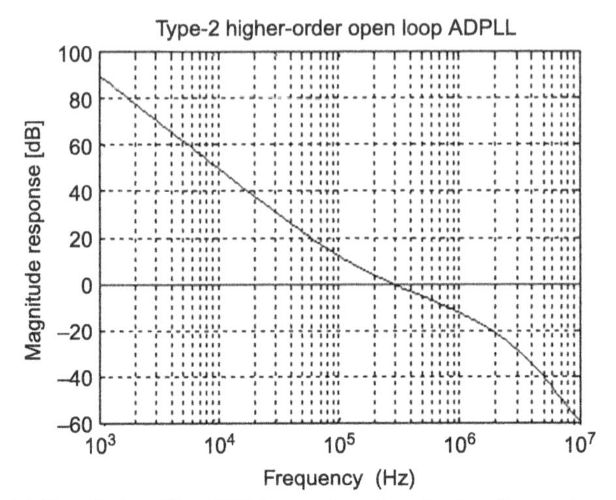

(a) Magnitude of the ADPLL open-loop transfer function vs. frequency.

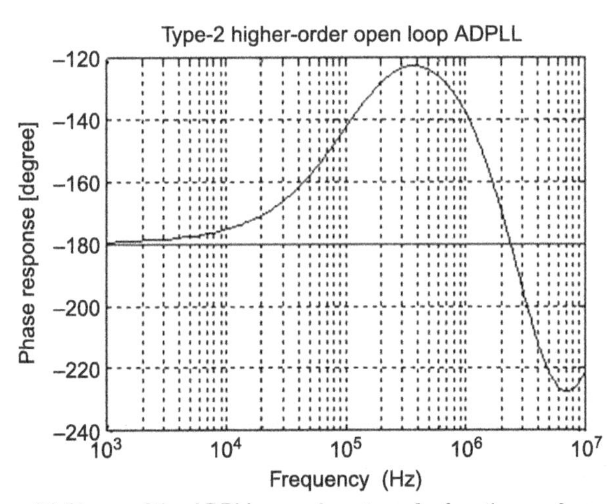

(b) Phase of the ADPLL open-loop transfer function vs. frequency.

Figure 4.13 ADPLL open-loop transfer function with $f_V = 62.001$ GHz, $f_R = 100$ MHz, $M = 32$, LF: $\alpha = 0.5$, $\rho = 1/2^8$, and $\lambda_i = 0.25, 0.25, 0.25, 0.25$.

4.6 NOISE AND ERROR SOURCES

A linear ADPLL model including phase noise sources is shown in Figure 4.14. The variable $\Phi_{n,R}$ is the phase noise of the external FREF. Its transfer function is the same as the closed-loop transfer function of Eq. (4.20). The TDC and DCO blocks are the only two places where

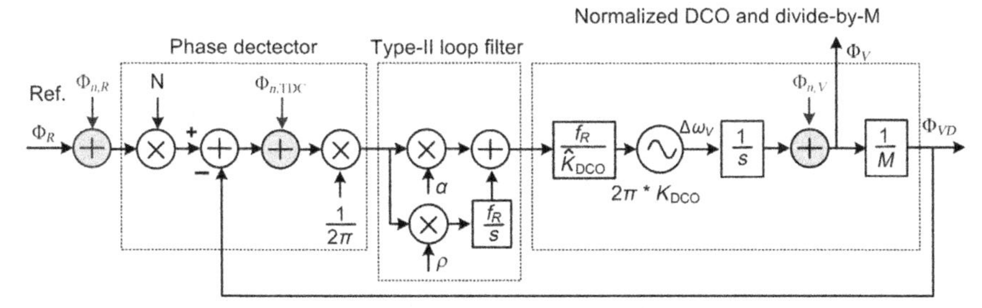

Figure 4.14 Linearized equivalent *s*-domain model of the ADPLL with noise sources.

noise could be injected to the system internally. Due to its digital nature, the rest of the system is free from time or amplitude perturbations.

The DCO phase noise $\Phi_{n,V}$ undergoes high-pass filtering by the loop. Its closed-loop transfer function is

$$H_{\mathrm{cl},V}(s) = \frac{\Phi_V}{\Phi_{n,V}} = \frac{1}{1 + H_{\mathrm{ol}}(s)} = \frac{s^2}{s^2 + \frac{\alpha f_R}{M} s + \frac{\rho}{M} f_R^2}, \tag{4.26}$$

indicating that the DCO noise has a high-pass characteristic and that it dominates the PLL phase noise outside of the loop bandwidth. At low-frequency offsets (i.e., within the loop bandwidth), the type II loop suppresses phase noise at 40 dB/decade in frequency.

The second internal noise source, $\Phi_{n,\mathrm{TDC}}$, arises from the TDC operation of calculating the fractional part of the variable phase. Even though the TDC is a digital circuit, the FREF and CKV inputs are continuous in the time domain. The TDC error has several components: quantization, linearity, and randomness due to thermal effects. Quantization noise (governed by Eq. 4.27) is often the major contributor [10],

$$\mathcal{L} = \frac{(2\pi)^2}{12} \left(\frac{\Delta t_{\mathrm{res}}}{M \cdot T_V} \right)^2 \frac{1}{f_R}. \tag{4.27}$$

In Eq. (4.27), Δt_{res} is the time resolution of the TDC, and $M \cdot T_V$ is the variable clock period at the TDC input. The factor M is due to the divider in the feedback loop. The closed-loop transfer function of the TDC noise can be expressed as

$$H_{\mathrm{cl}}(s) = \frac{\Phi_V}{\Phi_{n,\mathrm{TDC}}} = \frac{M \cdot H_{\mathrm{ol}}(s)}{1 + H_{\mathrm{ol}}(s)} = \frac{\alpha f_R s + \rho f_R^2}{s^2 + \frac{\alpha f_R}{M} s + \frac{\rho}{M} f_R^2}, \tag{4.28}$$

which is a low-pass response with a gain factor of M within the loop bandwidth. Therefore, the phase noise at the ADPLL RF output due to the TDC quantization noise is simply

$$\mathcal{L} = \frac{(2\pi)^2}{12}\left(\frac{\Delta t_{res}}{T_V}\right)^2 \frac{1}{f_R}, \tag{4.29}$$

within the loop bandwidth.

For example, if the TDC time resolution Δt_{res} is 12 ps, the reference clock f_R is chosen at 40 MHz, and the period of the variable clock T_V is 16.7 ps (i.e., $f_V = 60$ GHz), then the phase noise spectrum at the ADPLL RF output due to the TDC is −74 dBc/Hz. The noise of the external reference also sees a low-pass response; the total in-band phase noise at the RF output is the sum of the TDC noise and the reference noise.

Aside from the above three noise sources, the finite frequency resolution of the DCO also contributes phase noise to the ADPLL output. It should be kept much lower (e.g., 6 dB lower) than the natural phase noise of the DCO in order to be negligible. Similar to Eq. (4.27), the phase noise due to DCO frequency quantization is

$$\mathcal{L} = \frac{1}{12}\left(\frac{\Delta f_{res}}{\Delta f}\right)^2 \frac{1}{f_R}\left(\sin c\frac{\Delta f}{f_R}\right)^2, \tag{4.30}$$

where the sin*c* function arises from the Fourier transform of the zero-order hold operation [1]. Equation 4.30 gives rise to the same 20-dB/decade roll-off in phase noise seen from up-converted thermal noise in the oscillator, except for the protective notches at the DCO input sampling rate and its multiples. A ΣΔ dithering of the LSB in the DCO tuning bank is often employed to further reduce the quantization noise due to the finite tuning step of the DCO. With ΣΔ dithering, the phase noise spectrum in Eq. (4.30) becomes

$$\mathcal{L} = \frac{1}{12}\left(\frac{\Delta f_{res}}{\Delta f}\right)^2 \frac{1}{f_{dth}}\left(2\sin\frac{\pi\Delta f}{f_{dth}}\right)^{2n}, \tag{4.31}$$

where the f_{dth} is the dithering rate, and n is the order of ΣΔ modulator.

For example, adding an extra 8-bit first-order ΣΔ dithering would bring the phase noise down by 34 dB at lower-frequency offsets according to Eq. (4.31). The reader should be aware the dithering process itself also produces a significant amount of additional phase noise. The simulated phase noise spectrum due to Δf_{res} frequency quantization for various

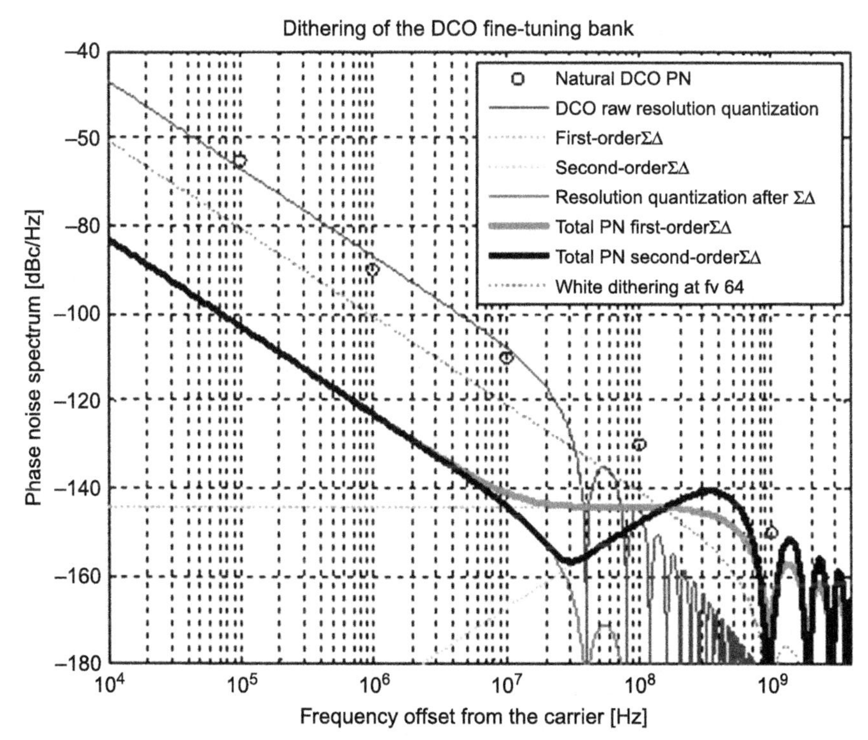

Figure 4.15 The simulated phase noise spectrum due to Δf_{res} frequency quantization for various dithering schemes ($\Delta f_{res} = 1$ MHz for a 60-GHz DCO without dithering).

dithering schemes is plotted in Figure 4.15. The free-running DCO phase noise curve is obtained from Spectre-RF™ simulation results for a 60-GHz DCO implemented in 65-nm CMOS. It can be seen that compared to white dithering at $f_V/64$ ($\sim$1 GHz), 8-bit $\Sigma\Delta$ dithering reduces the quantization noise more efficiently at lower-frequency offset, while shaping the noise at higher-frequency offsets. For first-order $\Sigma\Delta$ dithering, the total phase noise after dithering flattens off at the level of -145 dBc/Hz for offsets 20 MHz away from the carrier, dominating the noise floor at the synthesizer output. Employing second-order $\Sigma\Delta$ dithering can shape the DCO quantization noise to a higher-frequency offset (e.g., $\sim$200 MHz as shown in Figure 4.15), which is far from the frequency band of interest. Moreover, first-order $\Sigma\Delta$ suffers from a limit cycle intrinsically, and therefore, may introduce spurious tones at the ADPLL output.

4.7 BEHAVIORAL MODELING AND SIMULATION APPROACH

With an extremely high cost of several million dollars for a complete mask set in the latest CMOS technologies, it is imperative to fully validate the RF ADPLL prior to tapeout. A preferred simulator would allow seamless integration of RF, analog, and digital software at the top level. Successful use of the standard VHDL has been demonstrated for an ADPLL in Ref. [11]. To accommodate the greater role of software, new event-driven system simulators, such as SystemC and System-Verilog [12], are finding their ways into RF designs for the future. From the other direction, new mixed-signal-based simulators such as Verilog-AMS are being increasingly used to ease integration with SPICE-based simulators.

With the advent of fully digital frequency synthesizers for RF applications, a need has arisen to model and simulate RF circuits with the same simulation engine used for the digital back-end, which nowadays is likely to contain over a million gates. This way, complex interactions and performance of the entire SoC may be validated and verified prior to tapeout. Figure 4.8 offers some examples of these complex interactions:

1. Effect of the TDC resolution and nonlinearity on the close-in PLL phase noise and generated spurs.
2. Effect of the DCO phase noise on the PLL phase noise performance and generated spurs, especially when the PLL contains a higher-order digital loop filter and operates in fractional-N mode.
3. Effect of the DCO frequency resolution on the close-in phase noise of the PLL.
4. Effect of the $\Sigma\Delta$ DCO dithering on the far-out phase noise.

While SPICE-based simulation tools are extremely useful for small RF circuits containing several components (such as an RF oscillator or a low-noise amplifier), their long simulation times prevent investigation of larger circuits (such as an RF oscillator with a PLL loop and a transmitter or a receiver). Alternatively, the behavioral modeling and simulation environment based on a standard event-driven simulator (e.g., VHDL, Verilog) is well suited for digitally intensive SoCs with analog/RF circuitry. The main advantage of the single simulation engine at the top level is that it allows seamless integration of all hardware abstraction levels (such as behavioral, RTL, gate level) in a uniform environment. Extensive simulation and synthesis support by the VHDL or Verilog language makes it possible for a complex communication system to achieve the goal of "building what we simulate, and simulating what we build." Simulator performance, stability, multi-vendor support, a mature standard, and widespread use are all advantages of this environment.

4.8 SUMMARY

In this chapter we presented the fundamental operation of an ADPLL-based frequency synthesizer. The phase-domain all-digital phase correction mechanism closes the loop around an nDCO such that the oscillator's phase and frequency drift is corrected by means of the FREF. We also described the FREF retiming performed in such a way as to stochastically avoid metastability. A block diagram of the ADPLL described was presented in Figure 4.8. It contains only two non-ideal parameters in which it is not known exactly where distortion and noise could be introduced: the DCO gain (K_{DCO}) and the TDC resolution (Δt_{res}). The rest of the system is exact and is completely immune from any time- or amplitude-domain uncertainties and perturbations. In addition, the loop gain can be dynamically controlled via the gear-shifting techniques described in Section 4.4 to speed up frequency acquisition without sacrificing the phase noise performance after lock. For these reasons, the architecture described in this chapter appears to be highly competitive or even to exceed the performance of a conventional charge-pump PLL in low-gigahertz RF applications.

REFERENCES

[1] R.B. Staszewski, P.T. Balsara, All-Digital Frequency Synthesizer in Deep-Submicron CMOS, WILEY-Interscience, Hoboken, NJ, 2006.

[2] R.B. Staszewski, K. Muhammad, D. Leipold, C.-M. Hung, Y.-C. Ho, J.L. Wallberg, et al., All-digital TX frequency synthesizer and discrete-time receiver for Bluetooth radio in 130-nm CMOS, IEEE J. Solid-State Circuits 39 (12) (2004) 2278–2291.

[3] B. Nikolic, V.G. Oklobdzija, V. Stojanovic, W. Jia, J.K.-S. Chiu, M.-T. Leung, Improved sense-amplifier-based flip-flop: design and measurements, IEEE J. Solid-State Circuits 35 (6) (2000) 876–884.

[4] T.J. Gabara, G.J. Cyr, C.E. Stroud, Metastability of CMOS master/slave flip-flops, IEEE Trans. Circuits Syst. II 39 (10) (1992) 734–740.

[5] R.B. Staszewski, D. Leipold, P.T. Balsara, Just-in-time gain estimation of an RF digitally-controlled oscillator for digital direct frequency modulation, IEEE Trans. Circuits Syst. II 50 (11) (2003) 887–892.

[6] B. Razavi, Design of Monolithic Phase-Locked Loops and Clock Recovery Circuits: A Tutorial, in Monolithic Phase-Locked Loops and Clock Recovery Circuits: Theory and Design, IEEE Press, New York, NY, 1996.

[7] F.M. Gardner, Phaselock Techniques, Wiley, New York, NY, 1979.

[8] T.M. Almeida, M.S. Piedade, High performance analog and digital PLL design, Proc. IEEE Symp. Circuits Syst. 4 (1999) 394–397.
[9] R.B. Staszewski, G. Shriki, P.T. Balsara, All-digital PLL with ultrafast acquisition, in: Proceedings of the IEEE Asian Solid-State Circuits Conference, Taipei, Taiwan, sec. 11-7, Nov. 2005, pp. 289–292.
[10] R.B. Staszewski, K. Waheed, S. Vemulapalli, P. Vallur, M. Entezari, O.E. Eliezer, Elimination of spurious noise due to time-to-digita converter, in: Proceedings of IEEE Dallas Circuits and Systems Workshop, October 2009, pp. 1–4.
[11] R.B. Staszewski, C. Fernando, P.T. Balsara, Event-driven simulation and modeling of phase noise of an RF oscillator, IEEE Trans. Circuits Syst. I 52 (4) (2005) 723–733.
[12] T. Wen, T. Kwasniewski, Phase noise simulation and modeling of ADPLL by SystemVerilog, in: IEEE International Behavioral Modeling and Simulation Workshop (BMAS), September 2008, pp. 29–34.

CHAPTER 5

Millimeter-Wave Digitally Controlled Oscillator

Contents

At the heart of an ADPLL lies a digitally controlled oscillator (DCO) that does not require any analog tuning, but instead utilizes full digital control of resonating frequency. Millimeter (mm)-wave DCOs must be capable of tuning across a wide range (>10%) with fine frequency resolution (<1 MHz) and low phase noise. When used for a direct frequency modulation, linearity of the frequency tuning also becomes critical. Recent developments in mm-wave oscillator design presented in Chapter 3 can be applied to both voltage-controlled oscillators (VCOs) and DCOs. This chapter focuses on design issues unique to high-performance mm-wave DCOs, such as digital frequency tuning techniques. Careful design of the DCO interface is required when integrating it into a digitally intensive PLL. It will be described in the 60-GHz ADPLL example in Section 6.3. In this book, we use without loss of generality the design of a 60-GHz DCO intended for an FMCW radar application to illustrate principles and techniques behind the mm-wave DCO design. Before introducing

Millimeter-Wave Digitally Intensive Frequency Generation in CMOS.
DOI: http://dx.doi.org/10.1016/B978-0-12-802207-8.00005-8

these new circuit techniques, Section 5.1 reviews tuning methods for low-gigahertz DCOs and clarifies the design challenges when attempting to translate them to mm-wave frequencies. Then a reconfigurable passive resonator topology suitable for mm-wave DCOs is described in Section 5.2. Section 5.3 explains two fine-tuning techniques that achieve sub-megahertz frequency resolution in the 60-GHz band. The circuit implementations and experimental results for the 60-GHz DCO examples are elaborated in Section 5.4.

5.1 FROM LOW-GIGAHERTZ DCOS TO MM-WAVE DCOS

A conventional DCO is based on an LC-tank with a negative resistance to perpetuate the oscillation—just like the traditional VCO shown in Figure 5.1a. However, there is a significant difference in one of its components: instead of a continuously tuned varactor, the DCO now uses a large number of binary-controlled varactors (Figure 5.1b), as first proposed in Ref. [1]. Each varactor can be placed in either high or low capacitive state. A typical C–V curve of a MOS capacitor in deep-submicron CMOS is shown in Figure 5.2 with the high (off-state) and low (on-state) capacitive states highlighted. The composite varactor performs digital-to-capacitance conversion. Since the varactors are digitally controlled via a digital input word, and since the output clock at multi-gigahertz frequencies is of an acceptable digital waveform shape (the rise and fall times could be as fast as 20 ps), the loop around the DCO, which adjusts its phase and frequency, could now be *fully digital*, as first proposed in Ref. [2].

Figure 5.3 shows a simplified schematic of the DCO core that operates in the 3.2–4.0 GHz range. The DCO has not a single but actually three tuning word inputs to separately control the three varactor banks:

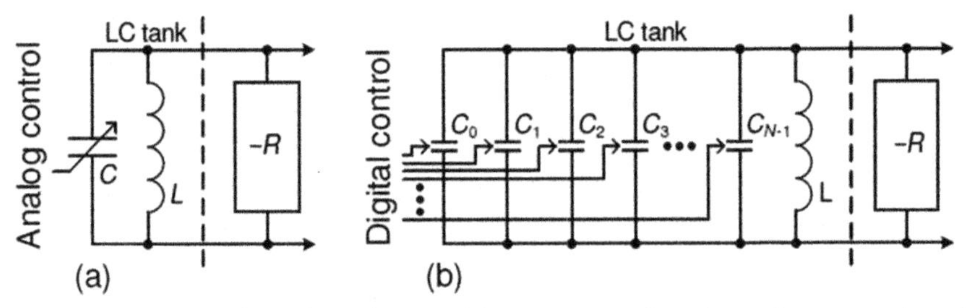

Figure 5.1 LC tank-based oscillators: (a) conventional with analog control and (b) with all-digital control.

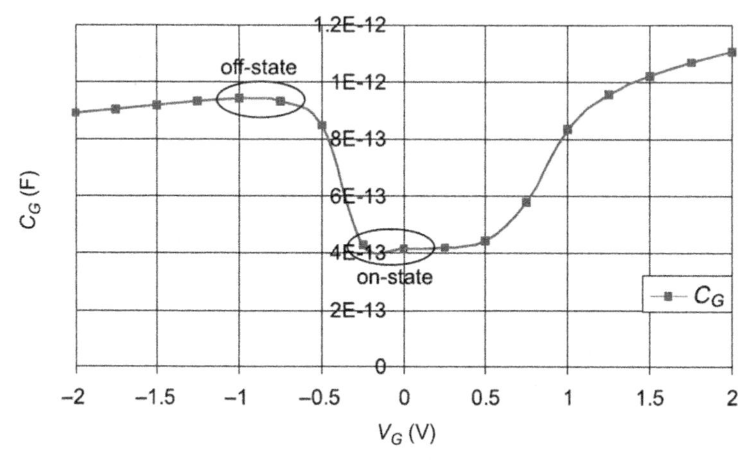

Figure 5.2 Gate capacitance vs. gate voltage of a measured PMOS varactor: 0.13-μm CMOS process, PPOLY/NWELL, inversion-type, single-contacted gate, $L = 0.5$ μm, $W = 0.6$ μm, $N = 8$ fingers × 12 × 2 [1].

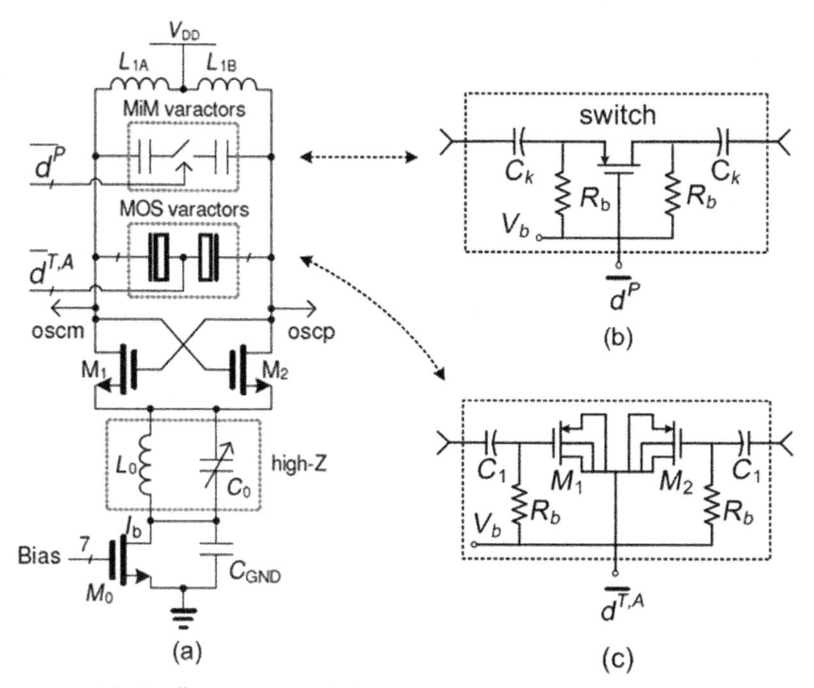

Figure 5.3 (a) Oscillator core and the varactor state driver array (GSM version example); (b) switched MIM capacitor; and (c) MOS varactor.

PVT centering (controlled by d^P), acquisition (controlled by d^A) and tracking (controlled by d^T). The PVT bank re-centers the DCO natural frequency to the middle of the selected frequency band. It has the largest frequency range since it has to cover all the frequency bands and margin for the oscillator variability. The acquisition bank performs channel selection by quickly settling to the neighborhood of the desired frequency. The tracking bank is the one actually used to track the oscillator drift during the transmission or reception. The capacitor banks are built using MIM and MOS varactors, as show in Figure 5.3b and c, respectively. The ADPLL quickly transverses these three varactor banks with progressively finer-frequency steps during locking (GSM example: 4 MHz, 200 kHz and 12 kHz, respectively). The oscillator phase noise is controlled by the dissipated current, which is established by the seven-bit "bias" control. In order to avoid biasing currents, the M_0 transistor array operates in linear region instead of in saturation.

Unfortunately, the fine frequency control used in low-gigahertz DCOs is not sufficient for mm-wave DCOs. The finest varactor step size made possible by the fine lithography is on the order of 40 aF in nanoscale CMOS, which corresponds to 12-kHz frequency step size at the 2-GHz DCO output. For an LC-tank resonating at 60-GHz band, typical tank inductance (L_0) and capacitance (C_0) values suitable for an oscillator implementation are 90 pH and 70 fF, respectively. An inductance smaller than 90 pH will have a lower-quality factor when implemented as a top-metal loop in most CMOS technologies. According to Eq. (3.2), the capacitance change (ΔC_0) required for a certain tuning step (Δf_0) is

$$\Delta C_0 = -\frac{\Delta\omega_0}{\omega_0}\cdot 2C_0 = -\frac{2\text{ MHz}}{60\text{ GHz}}\times(2\times 70\text{ fF}) = -4.7\text{ aF}. \qquad (5.1)$$

Thus, a fine-tuning step of 2 MHz for a 60-GHz carrier requires a capacitance change (ΔC_0) as small as 4.7 aF, which is 1/10th the 40-aF value of a minimum-sized NMOS varactor in nanoscale CMOS.

Besides the poor frequency tuning resolution, the oscillator tuning range (reflected in C_{max}/C_{min}) and the phase noise performance (largely determined by LC-tank Q-factor according to Eq. (3.1)) will both degrade dramatically when attempting to translate the DCO circuit in Figure 5.3 to the mm-wave bands. The Q-factors of an on-chip transmission line (TL), inductor and monolithic capacitors (including MOS varactors and backend metal capacitors) at mm-wave frequencies are compared in Ref. [3], and it is shown that both the TL and inductor offer much higher Q-factors than varactors (e.g., 20 for a 100-pH inductor versus 5 for a 50-fF capacitor at 60 GHz). Recall the analysis in Section 3.1: a tank Q-factor better than 10 is required in order to realize an LC oscillator with PN of −90 dBc/Hz at 1-MHz offset from a 60-GHz carrier. Consequently, the oscillator tuning range is limited to approximately 6% in order to achieve the tank Q-factor of 10 when varactor tuning is employed.

It is very challenging to design a wideband mm-wave DCO with a tuning range above 10% and a fine tuning step less than 1 MHz. Thus, it must come as no surprise that only two DCOs operating above 50 GHz have been reported prior to our work [4,5], and none has achieved 10% tuning range and sub-megahertz tuning steps simultaneously. Minimum-size switched MoM capacitors employed for fine tuning in Ref. [4] realized a frequency tuning step of 1.8 MHz for a 53-GHz DCO. However, the series parasitic capacitance of the MOS switch and the interconnections within the capacitor bank are much higher than the minimum MoM capacitance, which affects precision and matching.

In addition, both DCOs in Refs. [4,5] were reported as a standalone circuit building block and none of them were integrated into a PLL. Other tuning methods discussed in Chapter 3, such as inductive tuning using switched inductors [6] and coupled resonators [7], provide only coarse tuning with a very large step size (e.g., 1 GHz/bit) and must be augmented by switched capacitors to form a DCO. Consequently, those coarse-tuning and fine-tuning banks have different temperature coefficients due to their different frequency tuning mechanisms, complicating the calibration of K_{DCO} when used in a PLL. Thus, new circuit techniques are required to realize a DCO operating at mm-wave frequency with sufficient tuning resolution.

5.2 RECONFIGURABLE RESONATOR WITH DISTRIBUTED METAL CAPACITORS

To meet the stringent DCO performance requirements (e.g., >10% tuning range, 2-MHz raw resolution, PN at 1-MHz offset better than −90 dBc/Hz for a 60-GHz DCO) we propose a digitally reconfigurable passive resonator with distributed metal capacitors for application in mm-wave DCOs. The digitally reconfigurable resonator consists of a TL, inductor, or transformer, and pairs of metal shield strips located beneath the resonator and distributed along its major dimension in various metal layers, thereby exploiting the advantage of multiple metal layers available in nanoscale CMOS

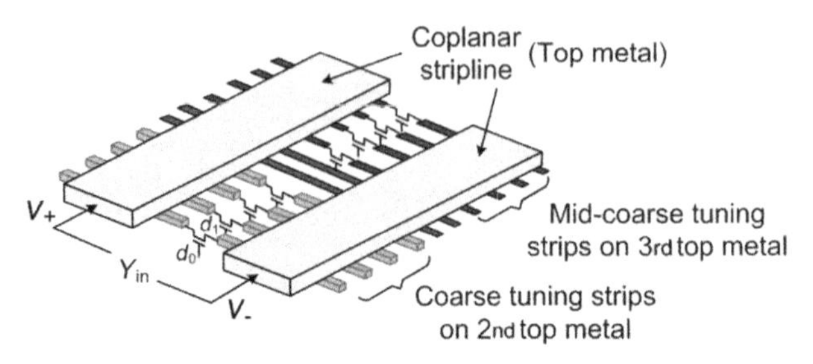

Figure 5.4 3D view of the digitally controlled TL.

technologies. Figure 5.4 illustrates the 3D view of a digitally controlled TL. Each metal shield strip pair is connected to a MOS switch driven by a digital tuning control. Activating the switch alters the capacitive load on the resonator and introduces a distinct phase shift in the DCO loop that changes the oscillator frequency. The inductive part of the passive resonator is implemented in thick top metal to reduce the conductor losses and to minimize loss from the silicon substrate. The digital tuning banks are distributed along the length of the resonator. The tuning elements are connected via the signal path of the resonator in top metal without any additional interconnecting wires, which eliminates complex (and mostly parasitic) wiring and optimizes the overall performance.

The switched-metal capacitors comprising the coarse-tuning and fine-tuning cells share the same temperature coefficient, which simplifies the calibration procedure when used in an ADPLL, and especially when it undergoes a direct frequency modulation. Electromagnetic (EM) simulations are required for the entire resonator structure to capture the distributed LC effects, including all metal switched-capacitor pairs and the inductor/TL/transformer as part of the DCO design procedure. The tank losses and DCO tuning characteristics can only be ascertained fairly accurately via EM simulations of the entire resonator physical layout (including the unwanted capacitive coupling between adjacent unit-tuning cells). The MOS switches cannot be included in an EM simulation, but they are added in subsequent circuit simulations of the DCO to analyze the tuning linearity and the effect of switch losses.

A unit-weighted tuning bank would require 2^{12} unit elements in order to achieve 6-GHz tuning range and 2-MHz raw frequency resolution. Capacitor mismatch is likely introduced when wiring the tuning capacitor banks together, as the interconnected parasitics are comparable in value to the tuning elements themselves. Furthermore, the mismatched parasitic capacitance introduces nonlinearity into the tuning characteristic (i.e., step size) of the bank. Therefore, the tuning capacitors are partitioned into three banks: a coarse-tuning bank (CB), a mid-coarse tuning bank (MB), and a fine-tuning bank (FB) in our 60-GHz DCO prototype. The design challenge for the CB is to maximize the tuning range assuming a (worst-case) tank Q-factor of 10. Wide tuning range is possible when a TL with a digitally controlled shield is employed in which the capacitance tuning step, ΔC, can be varied by placing metal strips on different metal layers. As for the FB, it should achieve ultra-fine-frequency resolution with minimal interconnection parasitics and linear tuning steps.

5.3 FINE-TUNING TECHNIQUES TO ACHIEVE HIGH-FREQUENCY RESOLUTION

Two fine-tuning techniques for high-resolution mm-wave DCOs are discussed in this section. The intention is to generate fine tuning steps without employing minimum-size structures so that the interconnection parasitics do not limit the frequency step size and uniformity. The two techniques incorporate the fine-tuning into either an inductor or transformer-based resonator. The fine-tuning bank is isolated from the TL-based coarse-tuning bank in the physical layout to minimize capacitive coupling between the different banks at mm-wave frequencies. The first technique exploits the variation in effect of a capacitor load on the tank resonant frequency with its position along the length of the resonator (i.e., variant tuning sensitivity). The second, transformer-coupled technique achieves fine-tuning via magnetic coupling between the primary and a switched metal capacitor bank placed beneath the secondary coil. Compared to a stand-alone inductor or transformer (without the digital tuning scheme), the degradation in its Q-factor due to fine-tuning is negligible (e.g., <0.5 from the total Q), since the embedded ΔC for each tuning bit is only approximately 4 aF ($\Delta f = 2$ MHz). Therefore, even when a small MOS switch is employed (e.g., $W/L = 1/0.1$ μm, on-resistance $R_{on} = 400\ \Omega$), the Q-factor of the series RC product for the switch is still above 3,300 in the 60-GHz band (as discussed in Section 3.1, $Q = \frac{1}{\omega \cdot \tau} = \frac{1}{2\pi f \cdot \Delta C \cdot R_{on}}$, in which τ is the RC time constant).

5.3.1 Inductor with Distributed Switched-C for Fine-Tuning

The inductor-based FB is illustrated conceptually in Figure 5.5a. An inductor is loaded differentially along its entire length by capacitors (C_L)

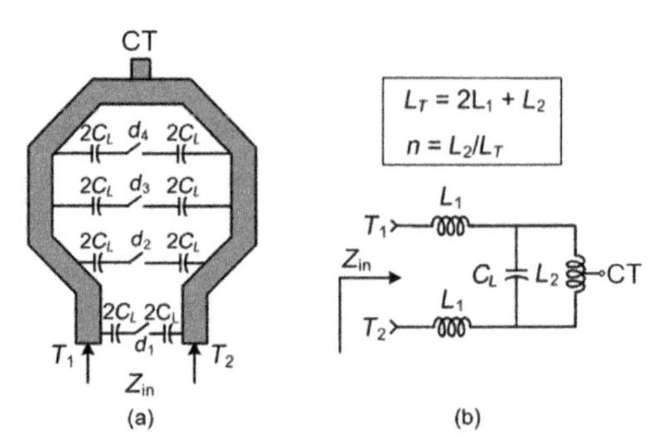

Figure 5.5 (a) Conceptual implementation of a distributed LC fine-tuning bank and (b) a simplified circuit model to illustrate the principle of tuning sensitivity attenuation [14].

from the differential inputs to the center-tap (CT). When the switch in series with C_L is ON, the inductor sees an increased capacitive load, which increases the differential input impedance (Z_{in}) of the tank as it approaches parallel resonance. Although the C_L array is unit-weighted, the sensitivity of Z_{in} to changes in C_L (i.e., $\partial Z_{\text{in}}/\partial C_L$) depends upon the position of the switched capacitor in the array. Placing C_L close to the center-tap introduces less phase shift compared to directly loading the inductor at the differential terminals connected to the oscillator core.

The reduced tuning sensitivity seen along the inductor length can be explained using the simplified circuit model shown in Figure 5.5b. The total loop inductance (L_T) between differential input terminals T_1 and T_2 is divided into three sections so that L_T consists of three parts (i.e., $L_T = L_1 + L_1 + L_2$). The mutual coupling between two L_1 segments and between L_1 and L_2 is negligible (mutual coupling factor is <0.1). The impedance Z_{in} seen across the terminals differentially when a single capacitor (C_L) is placed in parallel with L_2 is given by

$$Z_{\text{in}} = j\omega 2L_1 + \frac{1}{j\omega C_L} \,||\, j\omega L_2 = j\omega L_1\left[2 + \frac{(L_2/L_1)}{1-\omega^2 L_2 C_L}\right]. \tag{5.2}$$

Note that for L_2 approaching zero, $\partial Z_{\text{in}}/\partial C_L$ also approaches zero,

$$\frac{\partial Z_{\text{in}}}{\partial C_L} = j\omega\left[\frac{(\omega L_2)^2}{1-\omega^2 L_2 C_L}\right] = \begin{cases} 0, & L_2 \to 0 \\ j\omega \dfrac{(\omega L_T)^2}{1-\omega^2 L_T C_L}, & L_2 \to L_T \end{cases}, \tag{5.3}$$

and input impedance Z_{in} becomes insensitive to a change in C_L. Therefore, the inductor input impedance is desensitized to changes in capacitor C_L when it is placed close to the center-tap. On the other hand, when L_1 is zero and L_2 is maximum (i.e., $L_2 = L_T$; see Eq. (5.3)), $\partial Z_{in}/\partial C_L$ reaches its highest value possible.

To quantify the change in tuning sensitivity that is achieved by this method, we normalize $\partial Z_{in}/\partial C_L$ for an arbitrary L_2 to its maximum value (i.e., $\partial Z_{in}/\partial C_L$ at $L_2 = L_T$) and we name this factor the "normalized tuning sensitivity," α,

$$\alpha = \frac{\partial Z_{in}}{\partial C_L} \Bigg/ \frac{\partial Z_{in}}{\partial C_L}\Bigg|_{L_2=L_T} = \frac{n^2(1-\omega^2 L_T C_L)}{1-\omega^2 n L_T C_L} \approx n^2 (n \leq 1). \tag{5.4}$$

As can be seen from Eq. (5.4), α is proportional to the square of n, where $n = L_2/L_T$. Thus, the change in terminal impedance when capacitor C_L is added diminishes quickly if it is placed close to the inductor center-tap, rather than across the inductor input terminals (as in a conventional DCO).

Consider the implementation shown in Figure 5.5a. Four switched-capacitor unit cells are equally distributed along the inductor length and the minimum normalized tuning sensitivity α is 1/16 ($n = 1/4$). The value of α for each bit in the tunable inductor of Figure 5.5a is 1, 9/16, 1/4, and 1/16, respectively (in the sequence of inputs to CT). Thus, a fine-tuning bank of gradually reduced tuning sensitivity is formed. It is not necessary to distribute the tuning capacitors evenly along the inductor. A binary-weighted fine-tuning bank could also be obtained by sizing n as 1, $1/\sqrt{2}$, 1/2, $1/\sqrt{8}$, 1/4 in sequence, according to Eq. (5.4). When used in an ADPLL, nonlinearity in the tuning steps (i.e., K_{DCO}) could be compensated by adjusting the DCO gain normalization block ($1/\hat{K}_{DCO}$) as explained in Chapter 4, so that the PLL closed-loop bandwidth does not change.

To attenuate the tuning step further, more tuning cells can be placed closer to the inductor center-tap. For example, 16 equally distributed unit cells result in α of 1/256 for the cell located closest to the center tap. In practice, the maximum number of tuning cells that can be added is limited by the physical sizes of the inductor and MOS transistor used to switch the loading capacitance. For the 60-GHz LC tank where the outer dimension of the inductor is 80 μm × 60 μm, approximately 20 tuning capacitors can be added.

The above analysis neglects for simplicity the losses in the switched-capacitor unit cell. This is a fair assumption for a switched-capacitor structure used for fine tuning. The capacitance change ΔC is on the order of 0.1 fF, and even with the on-resistance of the MOS switch at

200 Ω, the Q-factor of this series RC network is still 132 at 60 GHz. Implementation of the inductor-based fine-tuning bank for a 60-GHz DCO prototype is described in Section 5.4, in which metal shielding strips connected to MOS switches are placed beneath the inductor, and are digitally controlled to act as the switched C_L shown in Figure 5.5a.

5.3.2 Transformer-Coupled Fine-Tuning Bank

A unit-weighted fine-tuning bank is desired for applications such as frequency ramp generation in FMCW radar applications. This section presents a fine-tuning technique that achieves a uniform frequency tuning step.

The tunable resonator consists of a transformer and a tunable load capacitor (C_L) connected to its secondary coil, as shown in Figure 5.6a. Resistor R_L models the losses of C_L. The primary coil of the transformer is connected directly to the oscillator core and a coarse switched-capacitor bank. The tunable transformer is analyzed as a one-port network for the admittance seen at the primary terminals (Y_{11}). The real part of Y_{11} (i.e., conductance G_{tran}) models the transformer losses. The imaginary part (i.e., inductive susceptance B_{tran}) in combination with the capacitive susceptance B_1 seen at the primary from the rest of oscillator determines the oscillation frequency, f_{osc}, such that $B_{\text{tran}} + B_1 = 0$. The coupling factor between the primary (L_p) and secondary (L_s) coils is k_m. When either C_L or R_L change, the change in admittance is reflected back to the primary coil and varies B_{tran}, thereby altering the oscillation frequency. The susceptance seen across the primary terminals (i.e., $L_{\text{eq}} = 1/j\omega B_{\text{tran}}$) can be varied in ultrafine steps even when the discrete tuning steps in C_L are moderate.

Numerical analysis is required to determine the Y_{11} and L_{eq} accurately since parasitic capacitances are difficult to determine precisely at high frequency, and capacitive effects are best investigated from simulating a particular case. However, some qualitative observations on the behavior of tunable transformer can be made from a simplified lumped-element circuit model shown in Figure 5.6b.

The parasitics to the substrate are neglected when analyzing the impedance transformation from the secondary to primary. The conductor losses r_s and r_p have a negligible effect on B_{tran} and are also ignored in the following analysis. Assuming that C_L is lossless (i.e., R_L very large), the equivalent inductance determined from Im[Y_{11}] is given by Eq. (5.5).

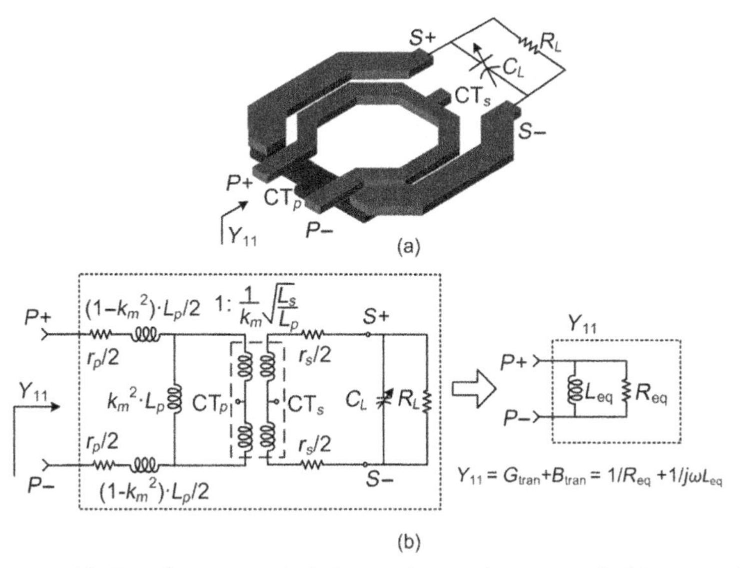

Figure 5.6 (a) Transformer-coupled fine-tuning technique and (b) a simplified lumped-circuit model for (a) [14].

When C_L is connected to L_s:

$$L_{\text{eq}}|_{R_L=\infty} = L_p\left(1 + k_m^2\frac{\omega^2 L_s C_L}{1-\omega^2 L_s C_L}\right) \xrightarrow{\omega^2 L_s C_L \ll 1} L_p(1 + k_m^2\omega^2 L_s C_L). \tag{5.5}$$

For comparison, the terminal inductance L_{eq} seen when C_L is connected directly to the primary coil is given by Eq. (5.6).

When C_L is connected to L_p

$$L_{\text{eq}}|_{R_L=\infty} = \frac{L_p}{1-\omega^2 L_p C_L} \xrightarrow{\omega^2 L_p C_L \ll 1} L_p(1 + \omega^2 L_p C_L). \tag{5.6}$$

Placing capacitor C_L across the secondary coil results in the same L_{eq} as when a capacitor of value equal to $C_L\cdot(k_m^2 L_s/L_p)$ is connected to the primary turn. In other words, the tuning sensitivity is attenuated by a factor of $k_m^2 L_s/L_p$, which can be much smaller than unity for a weakly coupled transformer (e.g., 0.01 for $k_m = 0.1$). Therefore, fine tuning of L_{eq} is possible using a capacitor bank with a moderate tuning step size. Furthermore, L_{eq} increases linearly with increasing C_L when the self-resonant frequency of the secondary coil $(1/\sqrt{L_s C_L})$ is much higher than the desired operating

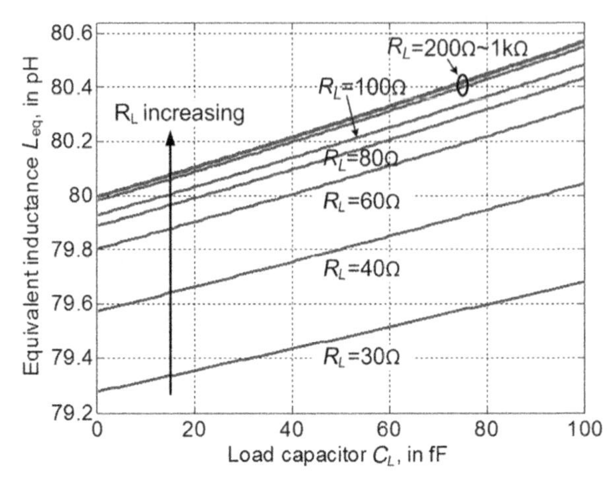

Figure 5.7 L_{eq} for variations in R_L and C_L at the secondary winding [14].

frequency, ω (i.e., $\omega^2 L_s C_L << 1$). Thus, a FB with a uniform tuning step can be achieved using a unit-weighted capacitor bank for C_L.

To investigate the effect of losses, variations in the primary admittance Y_{11} and L_{eq} (at 60 GHz) across C_L and R_L are simulated and plotted in Figure 5.7. A single-turn transformer with $L_p = 80$ pH and $L_s = 60$ pH at 60 GHz, and k_m of 0.2 is used in the simulation. As seen in Figure 5.7, the previous results derived assuming a lossless tuning capacitance C_L are still valid when R_L is larger than 200 Ω (i.e., L_{eq} is insensitive to R_L when $R_L > 200$ Ω), which is satisfied easily in a practical DCO. For a C_L of 10 fF in the FB, R_L ranges between 500 Ω and 1 kΩ at 60 GHz (i.e., Q-factor of 10–20).

The losses of the tunable transformer (i.e., $\mathrm{Re}(Y_{11}) = G_{tran}$) can be modeled by a resistor ($R_{eq} = 1/G_{tran}$) in parallel with L_{eq}. Figure 5.8 depicts the simulated R_{eq} for variations in C_L and R_L. It remains above 1 kΩ when R_L is higher than 200 Ω, indicating a negligible effect on the total tank Q-factor. On the other hand, for a small R_L (less than 100 Ω), L_{eq} depends not only on C_L but also increases rapidly with R_L, as shown in Figure 5.7. This attribute can also be employed to implement a variable inductor and a wide tuning range, but at the cost of poor tank Q-factor [8]. It is more desirable to vary C_L rather than R_L in order to achieve high DCO frequency resolution with less degradation in the tank Q-factor.

In order to obtain a linear frequency tuning characteristic, it was shown that capacitor C_L should satisfy the condition $\omega^2 L_s C_L << 1$. The capacitance attenuation factor can be increased by either reducing the

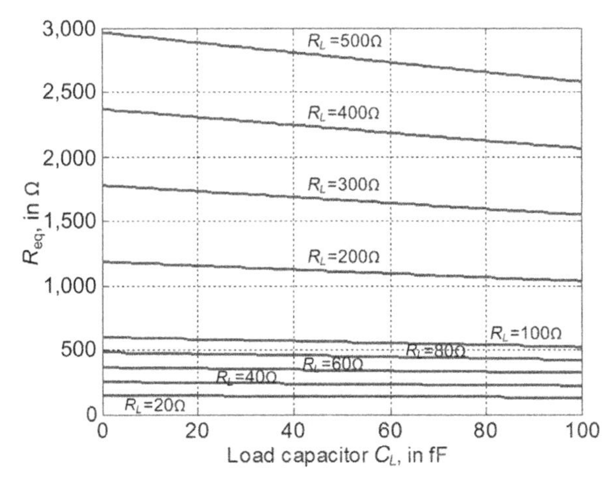

Figure 5.8 R_{eq} for variations in R_L and C_L at the secondary winding [14].

ratio of secondary to primary inductance (L_s/L_p), or by reducing the coupling coefficient k_m. However, k_m cannot be made lower than 0.1 because the transformer bandwidth also depends upon k_m [9], and it should be wide enough to cover the entire tuning range of the oscillator (e.g., 10% of 60 GHz for the DCO design examples shown in this chapter).

5.4 EXAMPLE IMPLEMENTATION OF 60-GHZ DCOS

Two 60-GHz high-resolution DCOs employing digitally reconfigurable passive resonators are described in this section. The first DCO, the L-DCO, is designed around an inductor-based fine-tuning bank, whereas the T-DCO employs a weakly coupled transformer to implement a unit-weighted fine-tuning bank. Both DCOs comprise three-stage segmented tuning: CB, FB, and MB that bridges the gap in step-size between CB and FB. Thermometer codes are employed to ensure monotonicity.

5.4.1 L-DCO

A simplified schematic of the 60-GHz L-DCO is shown in Figure 5.9. NMOS cross-coupled pair $M_{1,2}$ of 100-nm length and 18-μm width provides sufficient gain to sustain the oscillation. One oscillator output drives a divide-by-64 chain, while the other drives a single-ended RF output buffer for characterization. Three-dimensional views of the coarse- and fine-tuning structures are illustrated in Figures 5.10a and 5.11a, respectively.

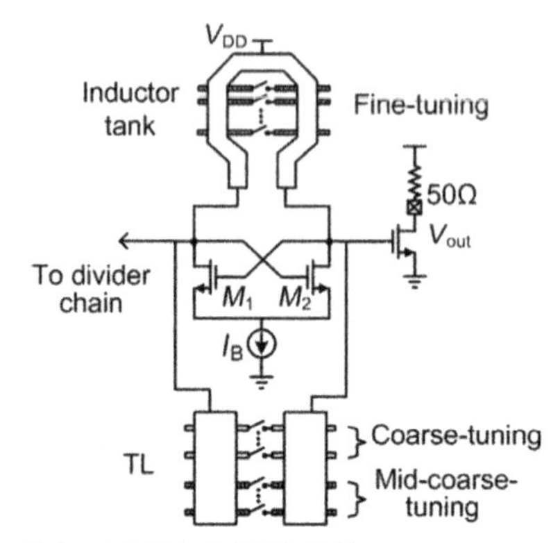

Figure 5.9 Schematic of the 60-GHz L-DCO [14].

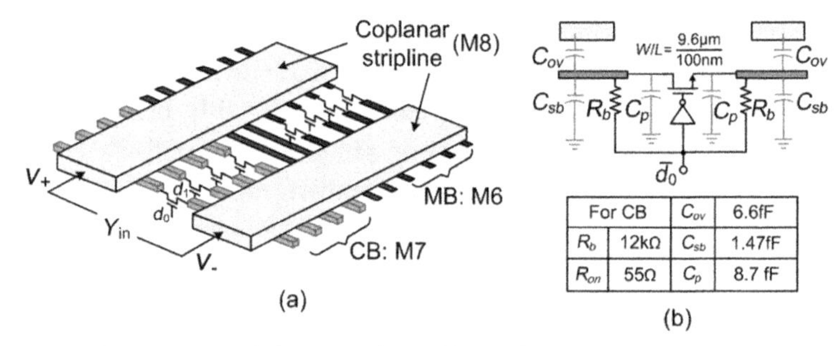

For CB		C_{ov}	6.6fF
R_b	12kΩ	C_{sb}	1.47fF
R_{on}	55Ω	C_p	8.7 fF

Figure 5.10 (a) 3D view of the reconfigurable TL for coarse- and mid-coarse-tuning and (b) detailed switch schematic, including parasitics [14].

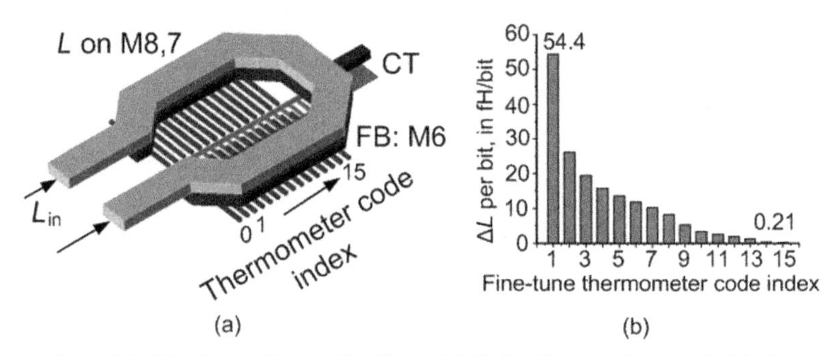

Figure 5.11 (a) 3D view of the distributed LC for fine-tuning and (b) simulated ΔL/bit [14].

A digitally controlled TL is used for both CB and MB as shown in Figure 5.10a. The 60-GHz oscillation signal runs across the TL of 34.5-μm width, 120-μm length and 39.6-μm spacing, which is implemented in 3-μm thick copper (M8 metal layer). Shorting metal strips (i.e., on M7 and M6) beneath the differential TL via NMOS switches increase the capacitance per unit length, thus reducing the wavelength ($\lambda = \frac{1}{f\sqrt{LC}}$) of the RF signal as described in Refs [10,11]. This increases the phase shift along the TL and reduces the tank resonant frequency. There is freedom in this design to dimension the metal strips and place them on different metal layers in order to obtain CB and MB with the desired frequency step ratio (e.g., 8). A 19-bit thermometer-coded CB and an 8-bit MB are implemented as shown in Figure 5.10a. Each bit in the CB (i.e., M7 strip) can introduce a ΔC of 1.2 fF, which corresponds to a frequency change of 315 MHz at 60 GHz, whereas the *MB* located on the (lower) M6 achieves a ΔC of 0.13 fF/bit ($\sim$1/8 of the CB step-size). The switchable capacitance ratio (C_{max}/C_{min}) of 1.6 provides over 6-GHz tuning range and a minimum tank Q-factor of 12 when all CB and MB switches are ON. Floating dummy strips added between the CB and MB help to minimize capacitive coupling between the two banks.

The design of MOS switches used in the CB (see Figure 5.10b) involves a trade-off between tuning range and Q-factor, which is exacerbated by the relatively high 60-GHz operating frequency [12]. The C_{ov} in Figure 5.10b represents the capacitance between the TL and the metal shield strip, and C_{sb} is the capacitance from metal shield strip to silicon substrate. When the NMOS switch is OFF, its impedance is dominated by the parasitic capacitance at the source/drain, C_p (e.g., at the source, C_p consists of source-bulk and source-gate capacitances, with bulk and gate acting as an AC-ground). The equivalent C_{min} is 2 fF as calculated from the data in Figure 5.10b. When the NMOS switch is ON, it operates in triode region with a resistance (R_{on}) of 55 Ω (transistor $W/L = 9.6/0.1$ μm). The equivalent C_{max} is 3.2 fF (i.e., $C_{max}/C_{min} = 1.6$) and the Q-factor drops to 15. Moreover, a salicide-blocked polyresistor of 12 kΩ (R_b in Figure 5.10b) biases the drain and source of the MOS switch at V_{DD} to ensure that the transistor is switched OFF, thereby reducing the parasitic capacitance due to the reverse biased drain—bulk junction [12]. In the ON state, it ensures the maximum gate-to-source (and drain) voltage for the smallest on-resistance. The MOS switch in MB is sized much smaller at $W/L = 3/0.1$ μm, as its ΔC is only an eighth of that of the CB.

A 3D view of the FB layout for the L-DCO is shown in Figure 5.11a. Equal-width metal shield strips are placed beneath the inductor on M6 (1-μm width and 1-μm spacing), which act as the capacitive load C_L to the inductor in Figure 5.5a. Shorting a metal strip pair with an NMOS switch increases the capacitive loading on the inductor. A thermometer code applied to the MOS switches from the differential inputs to the center-tap gives the simulated L_{in} change per bit (ΔL/bit) plotted in Figure 5.11b. The ΔL/bit is reduced progressively from 54.4 to 0.21 fH, corresponding to frequency steps of 23 MHz and 140 kHz at 60 GHz, respectively. The simulated input inductance L_{in} increases from 75.8861 to 76.0606 pH when the 15-bit thermometer code changes from all 0s to all 1s. The Q-factor of the fine-tuning bank is 20 and varies by ± 0.025 across the tuning range. The maximum achievable tuning step attenuation factor is 259 (54.4 fH/0.21 fH), which is very close to the theoretical prediction ($16^2 = 256$). Consequently, greater frequency resolution is achieved without exploiting minimum-size capacitors or varactors.

5.4.2 T-DCO

A simplified schematic of the T-DCO is shown in Figure 5.12a with the detailed implementation of the transformer-coupled FB illustrated in Figure 5.12b. Both primary and secondary coils are implemented in top metal (M8), with a (weak) coupling factor of 0.25 at low frequency. The primary coil is an M8-M7 stack to further reduce the conductor losses. The secondary coil is implemented in M8 only to maximize its self-resonance frequency since it will be loaded further by tunable capacitors.

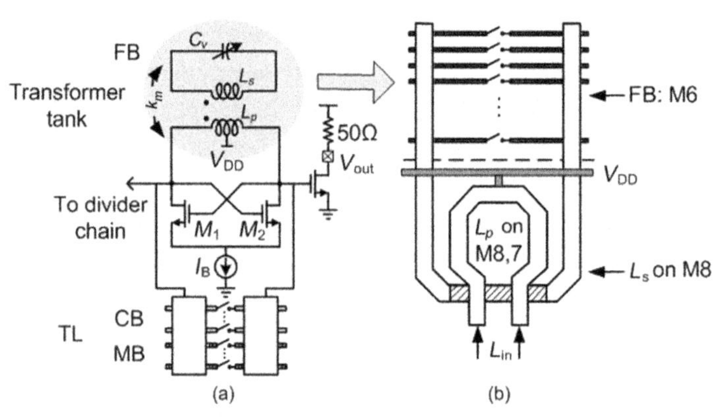

Figure 5.12 (a) Schematic of the 60-GHz T-DCO and (b) top layout view of T-DCO fine-tuning bank [14].

The variable capacitor load at the secondary coil is implemented as another digitally controlled differential TL with a much smaller tuning step compared to the one used for CB. The transformer and the TL-based tuning bank are co-designed using EM simulations to achieve the required Δf with a high Q-factor. Shorting each strip pair introduces a ΔC of 50 aF. The primary and secondary inductances of the transformer are 72.3 and 58.9 pH. The equivalent inductance change seen from the primary coil is 6 fH/bit, corresponding to approximately 2.5 MHz at 60 GHz. Finer frequency resolution can be obtained by resizing the shielding strips on the lower metal layers. Unwanted coupling between adjacent metal strips adds nonlinearity to the tuning curve. It is minimized by optimizing the width of the metal and the gap between adjacent strips with the aid of the EMX™ electromagnetic simulator [13]. The simulated Q-factor of the transformer-based FB is 16.5 in the 60-GHz band and varies by ± 0.03 across the tuning range.

5.4.3 Experiment Results

Two DCO prototypes verifying the proposed ideas were implemented in IBM's 90-nm CMOS on a 1.5 Ω-cm substrate. Die microphotographs of the DCO prototypes are shown in Figure 5.13 [14]. The core size for each design, including DCO, divide-by-64 stages, and the binary-to-thermometer decoder is $0.4 \times 0.4\ \text{mm}^2$. The L-DCO and T-DCO cores consume 10 and 12 mA, respectively, from a 1.2-V supply. The divide-by-64 chain and output buffers added for

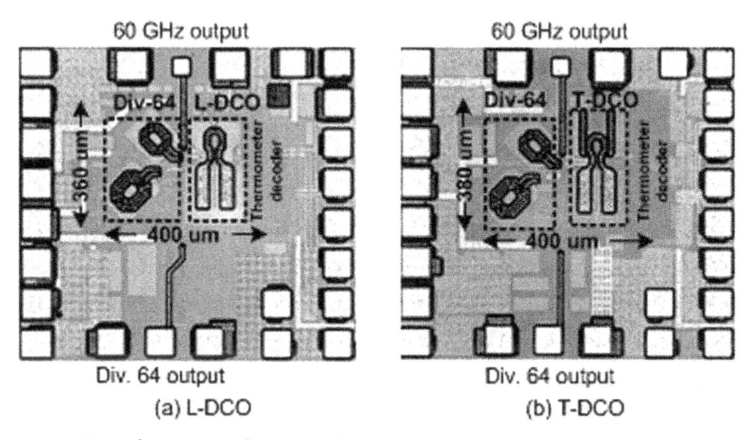

Figure 5.13 Test-chip microphotographs: (a) L-DCO and (b) T-DCO [14].

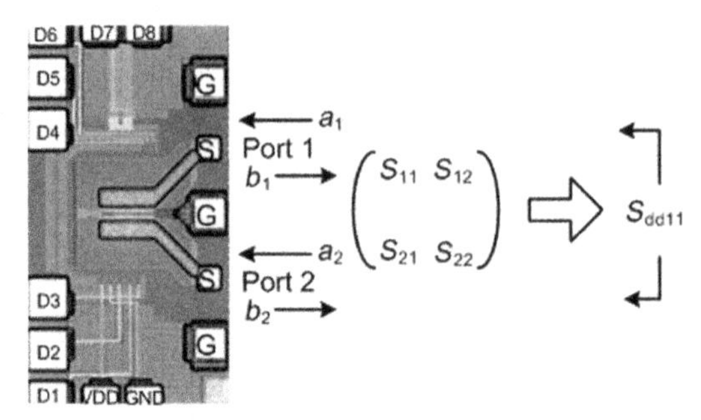

Figure 5.14 Die photo of the digitally controlled differential TL test structure.

characterization consume 18 mA in total. Five samples were measured to characterize each DCO.

5.4.3.1 Differential TL Test Structure

A test structure of the digitally controlled TL for the CB and MB was also fabricated and characterized separately. A die photo of the test structure is shown in Figure 5.14. Two-port S-parameters of the TL test structure including ground-signal-ground-signal-ground bondpads were measured from 50 GHz to 67 GHz for each digital control word using a vector network analyzer. The results were then converted to a one-port differential *S*-parameter (S_{dd11}) and plotted in Figure 5.15 for the extreme cases of all switches ON and all switches OFF. An extra phase shift is introduced at the TL input when all switches are ON. The simulated S_{dd11} of the test structure including bondpads is also plotted in Figure 5.15 for comparison. The measurement results agree well with the results predicted by EMX™ simulations.

5.4.3.2 L-DCO Measurement Results

The L-DCO has a measured tuning range from 56.15 GHz to 62.158 GHz. The measured and simulated coarse-tuning curves are plotted in Figure 5.16. The CB of the L-DCO realizes linear tuning at 312 MHz/bit with less than 12% variations. The linear tuning of MB at each CB code is plotted in Figure 5.17. More than 8% overlap between the adjacent frequency tuning curves in Figure 5.17 guarantees continuous tuning across the entire range. To accommodate PVT variations, the overlap ratio should be increased to 30% according to simulations. This can be achieved by adding more mid-coarse tuning strips to the TL structure.

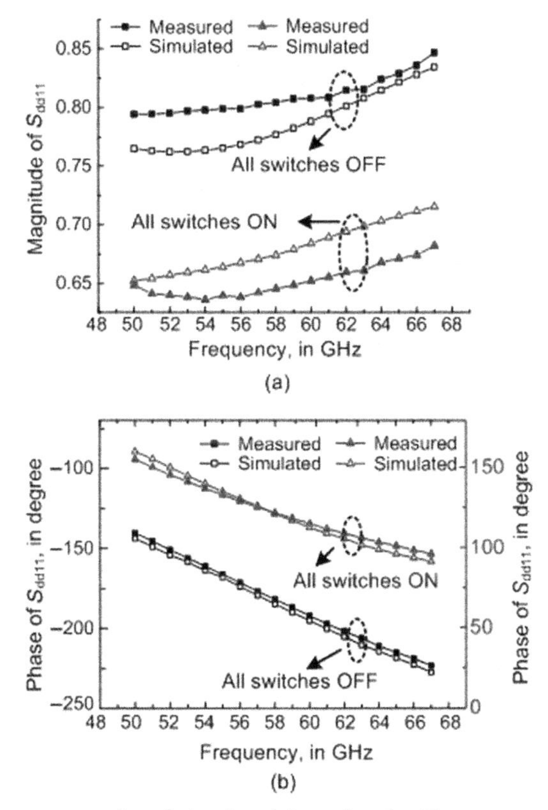

Figure 5.15 The measured and simulated S_{dd11} for the TL test structure: (a) magnitude and (b) phase.

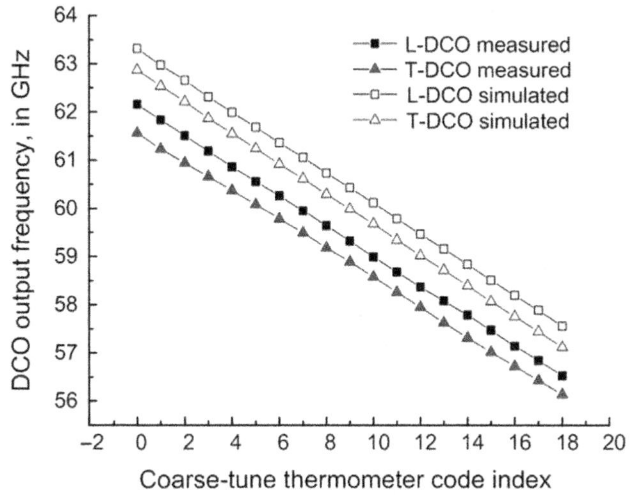

Figure 5.16 Coarse-tuning curves for the L- and T-DCOs [14].

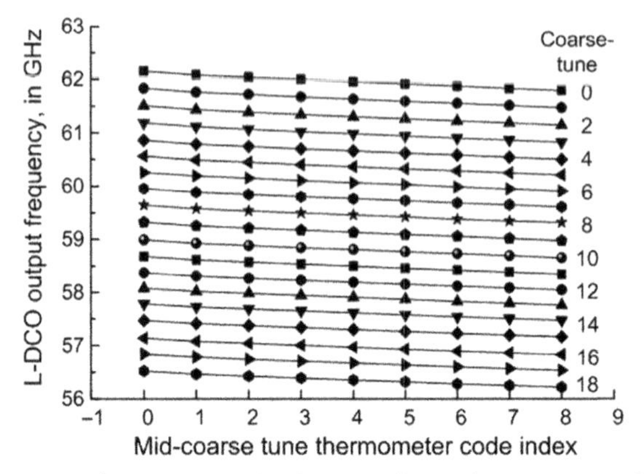

Figure 5.17 L-DCO mid-coarse-tune (MB) curves for each coarse-tune (CB) code [14].

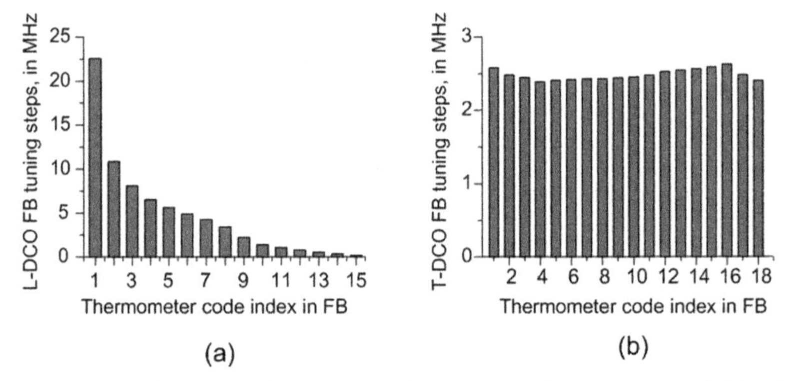

Figure 5.18 DCO frequency fine-tuning with respect to the input control: (a) L-DCO and (b) T-DCO [14].

The measured L-DCO fine-tuning frequency step for each bit in the FB (when CB = 8 and MB = 0) is shown in Figure 5.18a. A progressive reduction in step-size from 22.5 MHz for the first bit in the FB (i.e., farthest from the CT point), to 160 kHz for the last bit is observed. The worst case tuning range is 52.3 MHz for the FB (obtained at maximum CB and MB codes), which is over 30% larger than the tuning step size in the MB.

The RF output spectrum of the L-DCO oscillating at 60.864 GHz is shown in Figure 5.19. After subtracting cable and probe losses at 60.864 GHz (−13.5 dB loss), the measured output power is −3.4 dBm. The phase noise measured at the divide-by-64 output from a 951-MHz carrier is −127.83 dBc/Hz at 1-MHz offset, as shown in Figure 5.20.

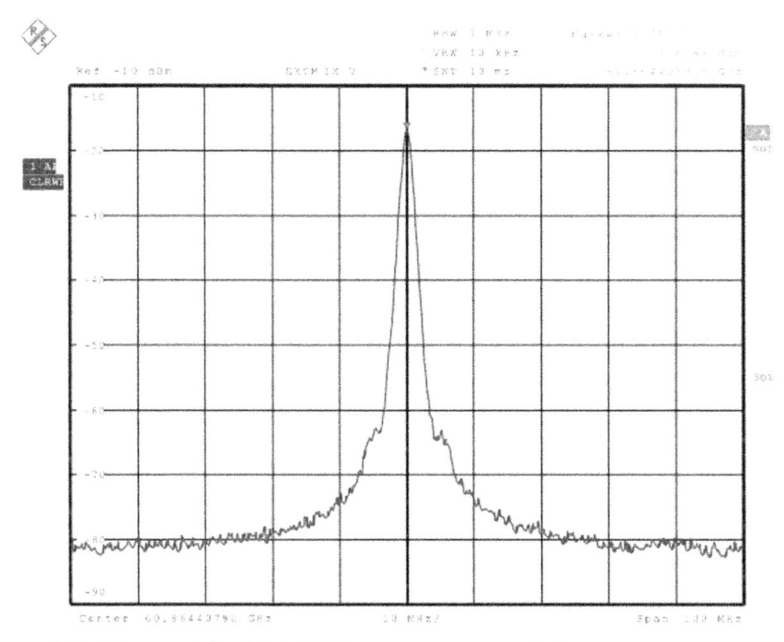

Figure 5.19 Measured 60-GHz L-DCO output spectrum [14].

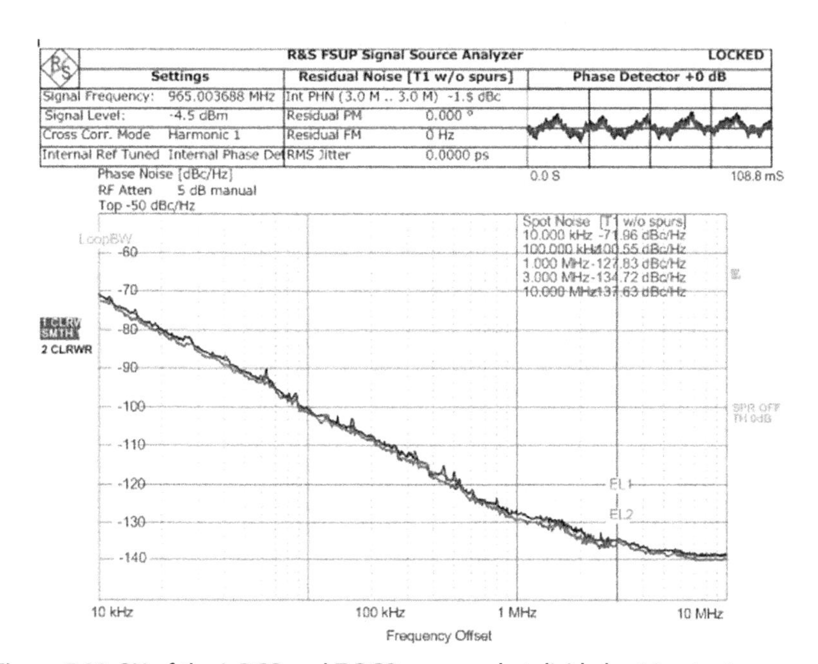

Figure 5.20 PN of the L-DCO and T-DCO measured at divide-by-64 output.

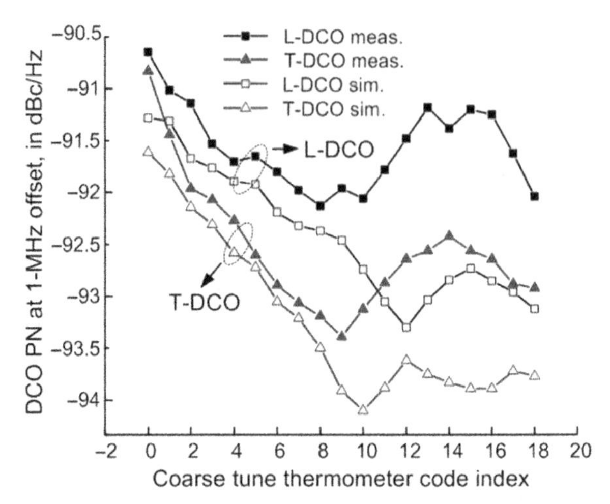

Figure 5.21 Measured PN of 60-GHz DCOs across tuning range ($\Delta f = 1$ MHz) [14].

The noise floor in Figure 5.20 is limited by the on-chip output buffer used for testing purposes. A separate measurement of the L-DCO output phase noise at 60.86 GHz (−92.5 dBc/Hz at 1-MHz offset) agrees well with the degradation expected after the division by 64 (theoretically $20\log_{10}(64) = 36.1$ dB). The measured L-DCO PN across the entire tuning range is better than −90.5 dBc/Hz at 1-MHz offset (Figure 5.21), which is just 1 dB poorer than predicted from simulations.

5.4.3.3 T-DCO Measurement Results

The measured T-DCO coarse-tuning characteristics are also plotted in Figure 5.16, which are comparable with the L-DCO. The T-DCO covers the range from 55.7 GHz to 61.56 GHz. The MB curve of the T-DCO is very close to that of the L-DCO, and thus it is not repeated here. The measured tuning step in the FB for each thermometer code is plotted in Figure 5.18b. The FB has a mean DCO tuning step of 2.5 MHz with less than 5% systematic mismatch, which could be digitally calibrated and compensated during frequency modulation, if necessary. One extra bit in the FB is reserved for $\Sigma\Delta$ dithering in the ADPLL. A first-order $\Sigma\Delta$ with 5 fractional bits would yield a frequency resolution of 78 kHz in the 60-GHz band. Phase noise at the divide-by-64 output of the T-DCO is also plotted in Figure 5.20. The PN across the entire tuning range is about 1 dB better than for the L-DCO (see Figure 5.21), and the output power is 0.5 dB higher due to the larger biasing current.

Table 5.1 Performance comparison of relevant state-of-the-art oscillators above 50 GHz

Ref.	Type	f_0 (GHz)	FTR (%)	Δf_{res}	PN (dBc/Hz)	P_{DC} (mW)	FoM_T (dBc/Hz)	FoM_{DT} (dBc/Hz)	CMOS
[5]	DCO	58.27 −63.83	9.3	2.3 bit	−90.1	10.6	−174.9	−162.2	90 nm
[4]	DCO	51.3 −53.3	4	10 bit 1.8 MHz	−116.5[a]	2.34	−179.2	−179.2	90 nm
[15]	VCO	53.2 −58.4	9.3	NA	−91	8.1	−176.6	NA	90 nm
[16]	VCO	64.8 −71	9.6	NA	−106[a]	5.4	−175.4	NA	65 nm SOI
[17]	VCO	54 −59.5	9.7	NA	−89	9.8	−173.9	NA	130 nm
L-DCO	DCO	56.22 −62.16	10	15.2 bit 160 kHz	−93	12	−177.9	−181.3	90 nm
T-DCO	DCO	55.83 −61.55	9.75	12 bit 2.5 MHz	−94	14	−177.9	−179.3	90 nm

[a]PN at $\Delta f = 10$ MHz; the rest are at 1 MHz.

The performance of the L- and T-DCOs are compared to other designs from the recent literature in Table 5.1 [4,5,15–17]. Only two DCOs above 50 GHz have been reported: Ref. [4] realized less than 4% tuning range and Ref. [5] obtained coarse-tuning (i.e., 370 MHz/bit) only. A commonly used oscillator figure-of-merit, $\mathrm{FoM_T}$, is defined in Eq. (5.7):

$$\mathrm{FoM_T} = \mathrm{PN} - 20\log_{10}(f_0/\Delta f \cdot \mathrm{FTR}/10) + 10\log_{10}(P_{\mathrm{DC}}/1\ \mathrm{mW}), \quad (5.7)$$

where PN stands for the phase noise at 1-MHz offset from carrier frequency f_0, FTR is the frequency tuning range in percent, and P_{DC} is the power dissipation of the oscillator. $\mathrm{FoM_T}$ includes the frequency tuning range but fails to account for the frequency resolution (Δf_{res}), which is crucial to DCOs. Therefore, we propose the DCO figure-of-merit, $\mathrm{FoM_{DT}}$ in Eq. (5.8), where N_{eff} is the effective number of tuning bits in the DCO as calculated by $\log_{10}(f_0 \cdot \mathrm{FTR}/100/\Delta f_{\mathrm{res}})$

$$\begin{aligned}\mathrm{FoM_{DT}} = \mathrm{PN} - 20\log_{10}(f_0/\Delta f \cdot \mathrm{FTR}/10 \cdot N_{\mathrm{eff}}/10) \\ + 10\log_{10}(P_{\mathrm{DC}}/1\ \mathrm{mW}).\end{aligned} \quad (5.8)$$

The L- and T-DCOs stand out with wide-tuning range and fine-frequency resolution for comparable PN performance, and achieve a higher $\mathrm{FoM_{DT}}$ compared to the other two DCOs operating at 60- and 50-GHz band, respectively.

5.5 SUMMARY

The digitally reconfigurable passive resonator with distributed metal capacitors in a mm-wave DCO achieves wide tuning range and small tuning steps simultaneously. Metal-oxide-metal capacitors are employed for all of the tuning banks, thus avoiding the use of low-*Q* MOS varactors. Two fine-tuning techniques have been demonstrated in 60-GHz DCO prototypes implemented in 90-nm CMOS. The distributed LC fine-tuning scheme in the L-DCO loads the inductor along its length capacitively, thereby modifying the tuning sensitivity of the LC tank. It realizes a frequency resolution better than 160 kHz at 60 GHz. The transformer-coupled fine-tuning scheme implemented in the T-DCO attenuates the effect of a change in capacitance via weak magnetic coupling so that practical capacitor values may be employed. The measured T-DCO fine-tuning step is 2.5 MHz. The T-DCO is not limited in its tuning arrangement by the physical size of the inductor (a limitation of the L-DCO) and a large number of fine-tuning bits can be implemented, as required. Moreover, a simple binary-to-thermometer decoder is sufficient for tuning word generation in an ADPLL with the T-DCO, since all three tuning banks are linear.

Both 60-GHz DCOs achieve a 6-GHz tuning range with more than 12-bit resolution and phase noise is better than −90.5 dBc/Hz at 1-MHz offset, which surpasses the performance of the previously reported designs. The passive resonator consisting of distributed switchable metal capacitors makes a wideband high-resolution DCO feasible at mm-wave frequencies. An all-digital PLL incorporating these high-performance DCOs would enable more mm-wave applications at a reduced cost for an implementation.

REFERENCES

[1] R.B. Staszewski, C.-M. Hung, D. Leipold, P.T. Balsara, A first multigigahertz digitally controlled oscillator for wireless applications, IEEE Trans. Microw. Theory Tech. 51 (11) (2003) 2154−2164.

[2] R.B. Staszewski, D. Leipold, K Muhammad, P.T. Balsara, Digitally controlled oscillator (DCO)-based architecture for RF frequency synthesis in a deep-submicrometer CMOS process, IEEE Trans. Circuits Syst. II 50 (11) (2003) 815−828.

[3] J.R. Long, Y. Zhao, W. Wu, M. Spirito, L. Vera, E. Gordon, Passive circuit technologies for mm-wave wireless systems on silicon, IEEE Trans. Circuits Syst. I Regul. Pap. 59 (8) (2012) 1680−1693.

[4] R. Genesi, F.M. De Paola, D. Manstretta, A 53 GHz DCO for mm-wave WPAN, Proc. IEEE Custom Integrated Circuits Conf., Sept. 2008, pp. 571−574.

[5] T. LaRocca, J. Liu, F. Wang, D. Murphy, F. Chang, CMOS digital controlled oscillator with embedded DiCAD resonator for 58−64 GHz linear frequency tuning and low phase noise, IEEE Int. Microwave Symp. Dig., June 2009, pp. 685−688.

[6] M. Demirkan, S.P. Bruss, R.R. Spencer, Design of wide tuning-range CMOS VCOs using switched coupled-inductors, IEEE J. Solid-State Circuits 43 (5) (2008) 1156−1163.

[7] J. Yin, H. C. Luong, A 57.5-to-90.1 GHz magnetically-tuned multi-mode CMOS VCO, Proc. IEEE Custom Integrated Circuits Conf., Sept. 2012, pp. 1−4.

[8] T. Lu, C. Yu, W. Chen, C. Wu, Wide tunning range 60 GHz VCO and 40 GHz DCO using single variable inductor, IEEE Trans. Circuits Syst. I Regul. Pap. 60 (2) (2013) 257−267.

[9] J.R. Long, Monolithic transformers for silicon RF IC design, IEEE J. Solid-State Circuits 35 (9) (2000) 1368−1382.

[10] T.S.D. Cheung, J.R. Long, K. Vaed, R. Volant, A. Chinthakindi, C.M. Schnabel, et al., Differentially-shielded monolithic inductors, Proc. IEEE Custom Integrated Circuits Conf., Sept. 2003, pp. 95−98.

[11] T. LaRocca, S.-W. Tam, D. Huang, Q. Gu, E. Socher, W. Hant, et al., Millimeter-wave CMOS digital controlled artificial dielectric differential mode transmission lines for reconfigurable ICs, IEEE Int. Microwave Symp. Dig., June 2008, pp. 181−184.

[12] H. Sjoland, Improved switched tuning of differential CMOS VCOs, IEEE Trans. Circuits Syst. II Analog Digit. Signal Process. 49 (5) (2002) 352–355.
[13] EMX™ User's Manual, Integrand Software, Inc., 2011.
[14] W. Wu, J.R. Long, R.B. Staszewski, High-resolution millimeter-wave digitally-controlled oscillators with reconfigurable passive resonators, IEEE J. Solid-State Circuits 48 (11) (2013) 2785–2794.
[15] L. Li, P. Reynaert, M.S.J. Steyaert, Design and Analysis of a 90 nm mm-wave oscillator using inductive-division LC tank, IEEE J. Solid-State Circuits 44 (7) (2009) 1950–1958.
[16] D.D. Kim, J. Kim, J.-O. Plouchart, C. Cho, W. Li, D. Lim, et al., A 70 GHz manufacturable complementary LC-VCO with 6.14 GHz tuning range in 65 nm SOI CMOS, IEEE Int. Solid-State Circuits Conf. Dig. Tech Papers, Feb. 2007, pp. 540–541.
[17] C. Cao, K.K. O, Millimeter-wave voltage-controlled oscillators in 0.13-μm CMOS technology, IEEE J. Solid-State Circuits 41 (6) (2006) 1297–1304.

CHAPTER 6

Application: A 60-GHz All-Digital PLL for FMCW Transmitter Applications

Contents

Millimeter-Wave Digitally Intensive Frequency Generation in CMOS.
DOI: http://dx.doi.org/10.1016/B978-0-12-802207-8.00006-X

The fundamentals of the all-digital PLL (ADPLL) architecture are described in Chapter 4. These basic understandings and design principles are employed here to design a mm-wave ADPLL. The digitally intensive nature provides flexibility, reconfigurability, and transfer-function precision, which are crucial in order to meet the diverse and strict requirements imposed by mm-wave applications. The ADPLL architecture and implementation is not restricted to a particular standard, and is applicable to frequency generation in mm-wave systems in general, for example, high data-rate communication at 60 GHz, 77/79-GHz automotive radar, and imaging at 94 GHz. This chapter uses a 60-GHz ADPLL as a design example to describe the multi-rate ADPLL-based frequency modulator architecture and to elaborate on the design techniques used to achieve high RF performance. The 60-GHz ADPLL was designed for a short-range FMCW radar transmitter [1]. The target specifications are outlined in Section 6.1. The multi-rate ADPLL-based frequency modulator architecture is described in Section 6.2 in detail, which features wideband frequency modulation capability and is especially attractive for the FMCW radar application. Aside from the high-performance DCO that is described already in Chapter 5, the design details of other key circuit building blocks are elaborated, including: the DCO interface, TDC, divider chain, and low-jitter reference slicer. Section 6.3 focuses on the DCO interface to the digital loop. High-frequency dividers are described in Section 6.4, while Section 6.5 covers the TDC design and calibration. Section 6.6 details the reference slicer and Section 6.7 discusses phase error glitch detection and removal. The top-level floor plan considerations for a mm-wave ADPLL are described in Section 6.8 followed by experimental results for the 60-GHz ADPLL prototype.

6.1 DESIGN SPECIFICATION

Millimeter (mm)-wave FMCW radars are utilized in automotive, security, and presence detection applications when high resolution is required [2,3]. For range-finding across a short distance indoors, the 60-GHz FMCW radar prototype aims for a resolution of 15 cm and an unambiguous range of 50 m. Its major specifications are listed in Table 6.1. A homodyne transceiver with triangular FM is a promising candidate for realizing a CMOS radar IC at low cost [4–7]. The frequency synthesizer in such radar ICs must provide a stable carrier of high spectral purity, and generate wideband, ultra-linear frequency modulation with programmable slope in order to obtain the required resolution. Its phase noise, chirp

Table 6.1 Indoor distance sensor performance

Range resolution	15 cm
Detection range	0.5 ~ 50 m
Operating frequency	60 GHz
Operation mode	Triangular frequency modulation

range (i.e., frequency modulation range), chirp slope (i.e., triangular modulation period, T_{mod}), and the linearity of the frequency sweep are the major design specifications.

6.1.1 Frequency Modulation Range (BW)

The operating principle of an FMCW radar transceiver is illustrated in Figure 1.3. For a short-range detection, a modulation range up to 1 GHz is required to obtain range resolution better than 15 cm, according to Eq. (1.1). Any nonlinearity in the frequency ramp results in range measurement errors, as the transmit signal is also used to detect the signal received from the target [8,9].

6.1.2 Frequency Modulation Period (T_{mod})

Chirp slope in an FMCW radar is defined as

$$k_{chirp} = \frac{BW}{T_{mod}/2}, \tag{6.1}$$

in which BW is the chirp range and T_{mod} is the chirp period, as shown in Figure 1.3. The chirp slope describes the rate of change in frequency with respect to the time traversing the triangular sweep. The resultant beat frequency f_b (i.e., intermediate frequency, IF) in a radar receiver is determined by

$$f_b = k_{chirp} \cdot \frac{2d}{c} \Rightarrow T_{mod} = \frac{4d}{c} \cdot \frac{BW}{f_b}, \tag{6.2}$$

in which $2d$ is the round trip distance between radar transmitter and receiver, and c is the speed of light. Since f_b is proportional to the round trip travel time, a fast chirp is desired to keep f_b above the flicker noise region of MOS devices (e.g., 100 kHz). By contrast, a slow chirp (e.g., T_{mod} of 10 ms) is required for high-resolution velocity detection, Δv, in a long-range scenario such as 77/79 GHz automotive radar ($\Delta v = \frac{c}{2} \frac{1}{T_{mod} \cdot f_c}$, where f_c is the center operating frequency) [10].

As listed in Table 6.1, the minimum and maximum detection ranges are 0.5 and 50 m, respectively. Thus, if f_b is 100 kHz for a target located 0.5 m away from the radar antenna, the highest value of f_b will be 10 MHz, which sets the minimum (Nyquist) sampling frequency of the analog-to-digital converter (ADC) in the receiver chain to 20 MHz. Further increasing the chirp slope pushes f_b away from the MOS transistor's flicker noise corner, but requires an ADC with a higher sampling rate, leading to increased power consumption. Taking this trade-off into account, a suitable range for T_{mod} falls between 50 μs and 2 ms for a sweep range of 1 GHz from Eq. (6.2).

6.1.3 Phase Noise

Phase noise influences the radar performance in several ways. The FMCW radar transmits and receives simultaneously, so the transmit signal will therefore leak to the receiver directly via the antennas and feeds, and also via the substrate when the transmitter and receiver are integrated on an IC. Due to the phase noise, the transmitter leakage could also mask targets since the targets are detected at frequency offsets from the carrier. The same is the case for strong nearby clutter.

Compared to a data communication system, the phase noise requirements on the LO are relaxed because the signal at the receiver is a delayed version of the transmitted signal. Thus, the phase noise of the two signals is correlated as seen from the expression for the phase noise difference, $\Delta\varphi(t)$

$$\Delta\varphi(t) = \varphi(t - t_d) - \varphi(t), \tag{6.3}$$

in which t_d is the time delay between the two signals mixing at the receiver. It is only the phase noise difference that plays a role. Without it, the beat tone would simply contain no phase noise if the transmitted and received signals contained identical phase noise at the same time. Taking into account the slopes of the output spectra due to phase noise, amplitude noise, and windowing of the sampled mixer output, it can be shown [7,8] that for adequate system design the window's power spectrum is the resolution-limiting factor, and therefore the effect of noise on the target resolution can be reduced by high sweep rates (i.e., high chirp slope).

Another important aspect about the oscillator noise is the resulting noise floor of the output spectrum, which is not influenced by correlation effects and directly determines the maximum signal-to-noise ratio that can be achieved in the system. For an FMCW transmitter using a PLL,

Table 6.2 FMCW synthesizer design specifications

Technology	CMOS 65 nm
Supply voltage	1.2 V
Power consumption	<50 mW
Chip area (core)	1 mm^2
CW mode	
Locking range	57 ~ 64 GHz
Settling time	<10 μs
Reference spur	<−60 dBc
In-band phase noise (at 10-kHz offset)	<−70 dBc/Hz
Out-of-band phase noise	−110 dBc/Hz at 10-MHz offset
Output power (after power amplifier)	+5 dBm
FMCW mode	
Triangular FM sweep range (BW)	>1 GHz
Triangular FM period (T_{mod})	50 μs to 2 ms
Triangular FM slope (k_{mod}) and required modulation clock (CKM)	k_{mod} >80 MHz/μs at CKM = 468 MHz k_{mod} ∈ [40, 80) MHz/μs at CKM = 234 MHz k_{mod} ∈ [20, 40) MHz/μs at CKM = 117 MHz k_{mod} <20 MHz/μs at CKM = 58.5 MHz
Sweep linearity: rms frequency error	<400 kHz

the phase noise floor for loop bandwidths on the order of several hundred kilohertz is dominated by the in-band phase noise. A typical phase noise requirement for a 60-GHz FMCW synthesizer is listed in Table 6.2, consistent with the published FMCW transceiver ICs in literature [4–7].

6.1.4 Frequency Sweep Linearity and Quantization Effect

The generated frequency sweep should be linear. Linearity is normally defined as the ratio of the maximum frequency deviation from the ideal chirp over the chirp bandwidth. Nonlinearities in the ramp result in spectral spreading of the beat frequency when transmit and receive waveforms are mixed. Thus, it is necessary to estimate the effect of modulation nonlinearity on the IF signal. A detailed derivation can be found in Ref. [8], from which the conclusions are: (1) even a small chirp nonlinearity would result in a considerable variation of the instantaneous frequency of the beat signals, (2) the final impact on the range resolution depends on the structure of the IF signal processing block in the receiver, the required magnitude of frequency deviation, and many other short-range radar parameters. Thus, the frequency error of the generated chirp should be kept to a minimum for optimum accuracy. State-of-the-art FMCW transceiver ICs implemented in CMOS technologies have achieved an rms frequency error of less than 400 kHz for various chirp slopes, resulting in a measured range error of less than 1%.

For an ADPLL-based FMCW generator, the frequency command word (FCW) is stepped up/down on each rising edge of the modulation clock (CKM), following the triangular FM trajectory. The modulation slope is controlled accurately by a fixed-point digital word with high precision, from which the step size of the FCW ramp is calculated and used to update the FCW in each CKM cycle. The DCO tuning characteristics need to be calibrated to achieve high-frequency sweep linearity. A faster CKM will produce a sweep that is closer to the ideal chirp at the expense of increased power dissipation. There is also a speed limitation set by the CMOS technology. For example, the maximum operating frequency for an 8-bit synchronous counter based on standard cells is less than 2 GHz in a 65-nm CMOS, at the SS process corner (slow NMOS and slow PMOS) with 1-V supply at 125°C. The effect of this quantization on the frequency sweep is analyzed in Ref. [8] to identify suitable values for modulation clock rate.

Table 6.2 summarizes the targeted specifications according to the above analysis. An operating frequency range of 7 GHz provides more than 1-GHz FM range with sufficient margin for PVT variations. If the 7-GHz full range is used for linear FM, the radar range resolution can be improved to 1.5 cm in theory. The chirp slope is programmable (50x) to accommodate the aforementioned trade-off between flicker noise and ADC sampling rate in the receiver. The CKM is provided from the divided DCO output with a division ratio of 2^n. Its optimum value for each chirp slope is also listed in Table 6.2. The synthesizer locking time and reference spur level achieved by state-of-the art 60-GHz PLLs from the literature are approximately 100 μs and −60 dBc, which are listed here for reference [11–13].

6.2 MULTI-RATE ADPLL-BASED FREQUENCY MODULATOR

Figure 6.1 depicts a block diagram of the mm-wave, ADPLL-based FM transmitter. The PLL topology is similar to the low-gigahertz ADPLL

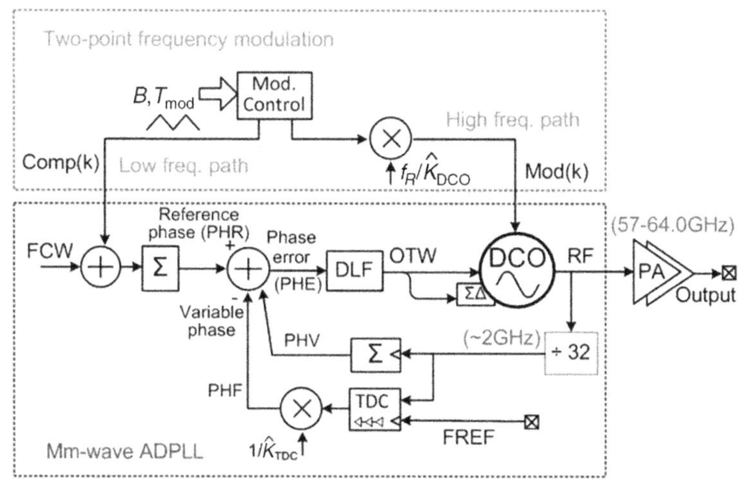

Figure 6.1 ADPLL with two-point FM.

discussed in Chapter 4. The difference is that the DCO is now oscillating in the mm-wave bands (e.g., 60 GHz) and a high-frequency divider chain is required in the feedback path to scale down the DCO output clock to a few gigahertz so that the phase detection can operate in the digital domain. Millimeter-wave DCO is the most critical building block in this architecture. The lack of high-performance DCOs previously hindered the development of ADPLLs operating in the mm-wave bands until the design techniques described in Chapter 5 were invented.

In Figure 6.1, a sufficiently slow wander of the DCO is corrected by the ADPLL negative feedback loop, which comprises a TDC to estimate the DCO phase, a FCW accumulator to calculate the reference phase, an arithmetic subtractor to calculate the phase error, and a digital loop filter to control the ADPLL bandwidth and its transfer function characteristics. The out-of-band phase noise is determined by the phase noise performance of the DCO, while the in-band phase noise is dominated by the TDC and the reference noise. Two-point frequency modulation is indicated in the upper part of Figure 6.1 to generate wideband FM. Let's first take a closer look at the 60-GHz ADPLL design before introducing the operation of two-point FM.

6.2.1 60-GHz ADPLL

The 60-GHz ADPLL operation is detailed in Figure 6.2. It operates synchronously in the phase domain. The DCO oscillates in the 60-GHz

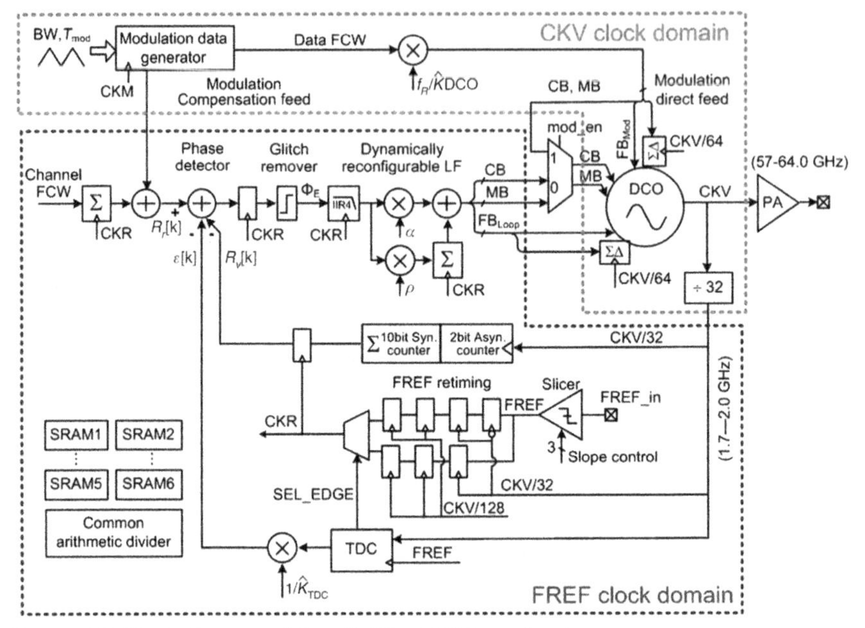

Figure 6.2 Block diagram of the 60-GHz ADPLL synthesizer.

band. It consists of three tuning banks to provide a 7-GHz tuning range with approximately 1-MHz raw frequency resolution, employing the transformer coupled fine-tuning techniques described in Section 5.3. A $\Sigma\Delta$ modulator is used to dither an LSB in the DCO fine tuning word at a high rate ($\sim$1 GHz) to further improve its frequency resolution down to 400 Hz. The divide-by-32 prescaler output CKV/32 at 2 GHz (f_V) features sharp rising and falling edges and serves as a digital clock for the rest of the loop.

The underlying frequency stability of the system is derived from an external frequency reference (FREF_in) generated by an external crystal oscillator ($f_R = 10\ldots100$ MHz). The FCW is defined as the desired frequency-division ratio f_V/f_R and is expressed in a fixed-point format such that the LSB of its integer part corresponds to the f_R reference frequency. It is input to the reference phase accumulator to establish the required operating frequency. The CKV/32 oversamples the frequency reference FREF to generate a retimed clock CKR as a synchronous system clock. The importance of reference clock retiming was previously explained in Section 4.2. To reduce metastability in FREF retiming, FREF is oversampled by both rising and falling edges of CKV/32 simultaneously, and an edge-selection signal derived from the TDC delay chain chooses the path furthest away from the metastable region.

The variable-phase R_v[k] signal in Figure 6.2 is determined by counting the number of rising clock transitions of the divided DCO oscillator clock (CKV/32). The reference phase R_r[k] is obtained by accumulating the FCW with every rising edge of the retimed FREF clock (CKR). The variable-phase R_v[k], together with the fractional correction ε[k], is subtracted from the reference phase R_r[k] in a synchronous arithmetic-phase detector. The ε[k] corrections by means of the TDC system increase the instantaneous phase resolution of the system to below the basic 2π radians of the variable phase. A simplified glitch removal circuit compares the absolute value of the Φ_E jump with a half-integer threshold to correct potential misalignment between R_v[k] and the ε[k] coming from the TDC.

The digital phase error Φ_E[k] is conditioned by a reconfigurable LF. The LF here is similar to the LF described in Section 4.4. It can be a proportional attenuator α, forming a type I loop for fast frequency/phase acquisition during the lock-in process. The LF can also be configured as type II to offer better filtering of oscillator noise within the loop bandwidth. Adding a fourth-order IIR filter to suppress the TDC and reference noise outside of the loop bandwidth leads to improvements in the overall phase noise performance. The LF employs the gearshift technique (see Section 4.4) to minimize the settling time by switching dynamically the loop coefficients during the frequency acquisition, in order to keep the phase noise as low as possible in the steady state. A 3-bit slope control in the FREF slicer reduces the reference spur amplitude at the cost of a slight increase in the in-band PN. The built-in DCO (K_{DCO}) and TDC gain (K_{TDC}) calibrations are performed automatically to ensure a wideband FM response. The TDC gain calibration will be discussed in Section 6.5, while the DCO gain linearization with be described in Chapter 7. Six 8-kbit SRAMs and other digital arithmetic blocks are also integrated on-chip to enable system debugging. These memories also supply look-up tables when applying wideband FM.

This digital phase-domain operation keeps the phase information in fixed-point numbers (after the necessary initial conversions from "analog" domains) that cannot be further corrupted by noise. Consequently, the phase detector can be realized simply as an arithmetic subtractor that performs an exact digital operation without generating reference spurs, which is not the case in a charge-pump PLL. The dynamic range of the phase error could be made arbitrarily large by increasing the word-length of the phase accumulators. This compares favorably to more conventional implementations, which typically are limited to only $\pm 2\pi$ of the comparison rate with a three-state phase/frequency detector [14].

6.2.2 Wideband Frequency Modulation

The ADPLL features a wideband FM capability, which can be realized as a two-point modulation scheme [15]. In Figure 6.1, one data path directly modulates a DCO, while the other path compensates the reference path and prevents the modulating signal from affecting the phase error. The former path has a high-pass characteristic, while the latter low-pass filters the signal. When both paths are combined (with a compensation for any time misalignments via digital algorithms, if necessary), an all-pass transfer function is realized. Therefore, the maximum modulation frequency is not limited by the PLL closed-loop bandwidth, and an ultrafast linear ramp can be synthesized.

Two-point FM has been demonstrated in numerous prototypes at low-gigahertz frequencies [15–18]. For the two-point modulation scheme to work properly, the modulating data must be normalized accurately to the DCO gain (K_{DCO}) in the direct modulation path (i.e., $f_R/\hat{K}_{\mathrm{DCO}}$ multiplier in Figures 6.1 and 6.2). If the normalization is exact (i.e., the estimate is accurate: $\hat{K}_{\mathrm{DCO}} = K_{\mathrm{DCO}}$), the modulating transfer function is flat from dc to $f_s/2$ in the z-domain, and has only a sinc-type response in the s-domain caused by the zero-order hold of the DCO interface. In order to synthesize a linear chirp of several gigahertz in range, multiple DCO tuning banks of various tuning step sizes (i.e., different K_{DCO}) are employed. A closed-loop DCO gain linearization algorithm (described in Chapter 7) compensates for the PVT variations of K_{DCO}, and the calibration data are stored in an SRAM look-up table. Upon modulation, a predistorted signal is applied in the data path of the DCO to obtain higher sweep linearity across a gigahertz modulation range.

The modulation data samples in this ADPLL can operate at a high rate (CKM clock in Figure 6.2) obtained by a low integer division of the variable DCO clock (CKV), which is independent of the phase detection rate at the re-sampled reference frequency (CKR). Therefore, the ADPLL-based frequency modulator architecture shown in Figure 6.2 is a multirate system. The CKM is programmed to be fast enough to overcome the effect of quantized frequency sweep on the IF signal in the receiver [8]. Being digital, the modulating paths have a clock-cycle precision in time. Thus, choosing a high rate for CKM reduces the instantaneous frequency error with respect to an ideal ramp in the synthesized FMCW signal, thereby improving the chirp linearity.

6.3 DCO INTERFACING

An improved version of the transformer-coupled high-resolution DCO presented in Chapter 5 is employed here at the heart of the 60-GHz ADPLL. The DCO operational principle and the design details of proof-of-concept 60-GHz DCO in 90-nm CMOS were described in Chapter 5. In this section, we focus on the improvements and the interfacing design considerations when used in an ADPLL. Figure 6.3 shows how the DCO could fit into the ADPLL system. Its raw OTW is partitioned into four data buses feeding four tuning banks: coarse-tuning (CB), mid-coarse-tuning (MB), fine-tuning (FB), and a high-speed fractional FB dithering bank with kilohertz-level resolution. Moreover, a separate fine-tuning (FB_{Mod}) is dedicated for modulation. The benefit of separating the fine-tuning for the loop from the modulation path will be explained shortly. These data buses are binary-to-thermometer encoded to digitally control the unit-weighted tuning elements in the tank, thereby ultimately setting the DCO to the desired frequency. The 60-GHz DCO drives an output power amplifier and the prescaler. The latter provides a 2-GHz signal for the TDC and the digital part of the ADPLL.

The schematic of the 60-GHz DCO together with its load are plotted in Figure 6.4a. The NMOS cross-coupled pair ($M_{1,2}$) sustains the oscillation. The DCO is segmented into three banks, each with a linear tuning

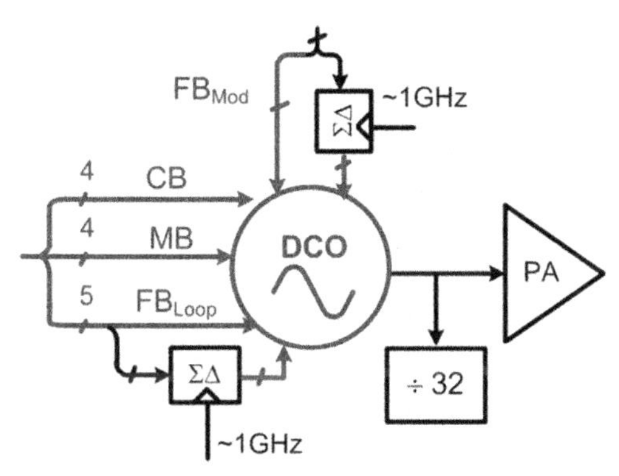

Figure 6.3 60-GHz DCO in the ADPLL system.

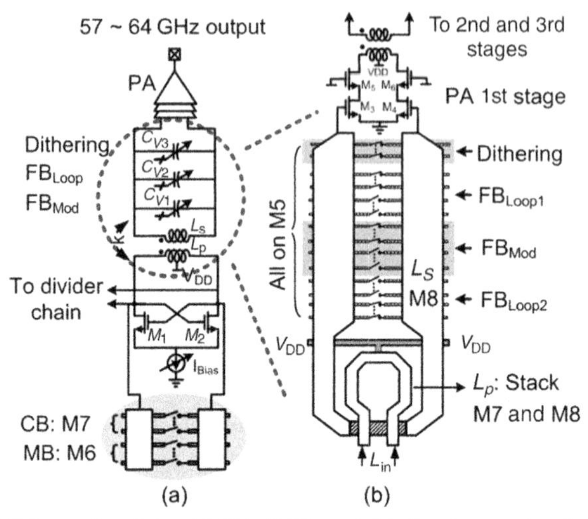

Figure 6.4 (a) Schematic of the 60-GHz DCO and (b) the fine-tuning bank.

characteristic: CB, FB, and MB. The MB bridges the gap in step-size between CB and FB. The CB and MB are integrated with the TL and configurable floating metal shield to form a compact, digitally controlled frequency-tuning scheme. A smaller tuning step is attained by placing metal strips on (lower) metal M6 compared to the coarse-tuning strips on metal M7. The fine-tuning bank (C_{V1-3}) is placed at the transformer secondary winding with a weak mutual coupling factor k_m of 0.28 to attenuate its frequency-tuning sensitivity by a factor greater than 10. The measured DCO oscillation range is from 56.4 to 63.4 GHz with a coarse-tuning resolution of 367 MHz/bit, mid-coarse-tuning of 35 MHz/bit, and fine-tuning of 1.64 MHz/bit. The latter is further enhanced through dithering.

6.3.1 Separate DCO Fine-Tuning Bank for CW and FM

To optimize the ADPLL operation in both continuous-wave (CW) and FM modes, the FB is split into two parts, as shown in Figure 6.4b; FB_{Mod} at the center of the TL is dedicated for frequency modulation, and FB_{Loop}, located above and below FB_{Mod}, is used to correct DCO phase/frequency wander in the loop at slow rates. In this way, only K_{DCO} of FB_{Mod} needs to be calibrated accurately (e.g., <5% mismatch) and applied to the $f_R/\hat{K}_{DCO}$ multiplier in the direct modulation path of Figure 6.1. The FB_{Loop} can tolerate a much larger tuning-step mismatch (e.g., 15%) and only a rough approximation of the DCO transfer function

K_{DCO} (e.g., 20%) is required to establish an acceptable range for the ADPLL loop bandwidth. The loop bandwidth affects mainly the settling time and noise rejection of the PLL, so a 5–25% variation would have minimal effect on the system performance [15]. Therefore, a K_{DCO} of 1 MHz/bit is used in the $1/\hat{K}_{DCO}$ multiplier in the loop, and the DCO gain ratios between different banks are scaled by 2^n (i.e., CB/FB = 32 and MB/FB = 16) to simplify multiplication to a right-bit-shift. Consequently, the system complexity and ADPLL loop delay are reduced, which improves the phase margin in wide bandwidth operation.

6.3.2 Decoder Mapping Algorithm for Device Matching

Device matching in the FB_{Mod} circuit is critical for distortion-free modulation. To improve device matching, dummy cells are normally added to the capacitor tuning bank in conventional DCOs. However, the need to minimize all parasitics does not permit the addition of dummy cells in a 60-GHz design. Moreover, the metal shield strips beneath the TL must be controlled monotonically to ensure an unambiguous K_{DCO} for each bit. The unavoidable coupling between adjacent strips makes K_{DCO} dependent on nearby states, which increases the tuning nonlinearity in FB. To overcome these problems, a novel decoding scheme for the FB is illustrated in Figure 6.5. By default, half of the switches in each part (FB_{Mod} and FB_{Loop}) are turned ON via re-centering of the fine-tuning bank after locking [19]. The switches in the lower half-part, FB_{Loop1}, are ON ("1") and in the upper half-part of FB_{Loop2} are OFF ("0"); both acting as dummies for FB_{Mod} in the center (state 0). When a small frequency drift upwards appears in the loop, the fine-tuning bank changes to state " + " (see Figure 6.5) in response to the positive phase error. To track even more frequency drift upwards, it then changes to state "+ +". The switches are turned ON in sequence shown in Figure 6.5. Enough "virtual" dummy switches are realized in this way for FB_{Mod} unless the frequency drift is so large that all the switches in FB_{Loop} are ON, which should not happen at all in a normal operation. A similar scenario for frequency drift downward is shown as state "-" and state "- -". Hence, less than 0.1% K_{DCO} variation in FB_{Mod} is measured at various FB_{Loop} settings, even without extra dummy cells.

To further improve the DCO's frequency resolution, a (programmable) first/second-order $\Sigma\Delta$ dithering at CKV/64 ($\sim$1 GHz) is implemented to reduce the DCO quantization phase noise to below -140 dBc/Hz at

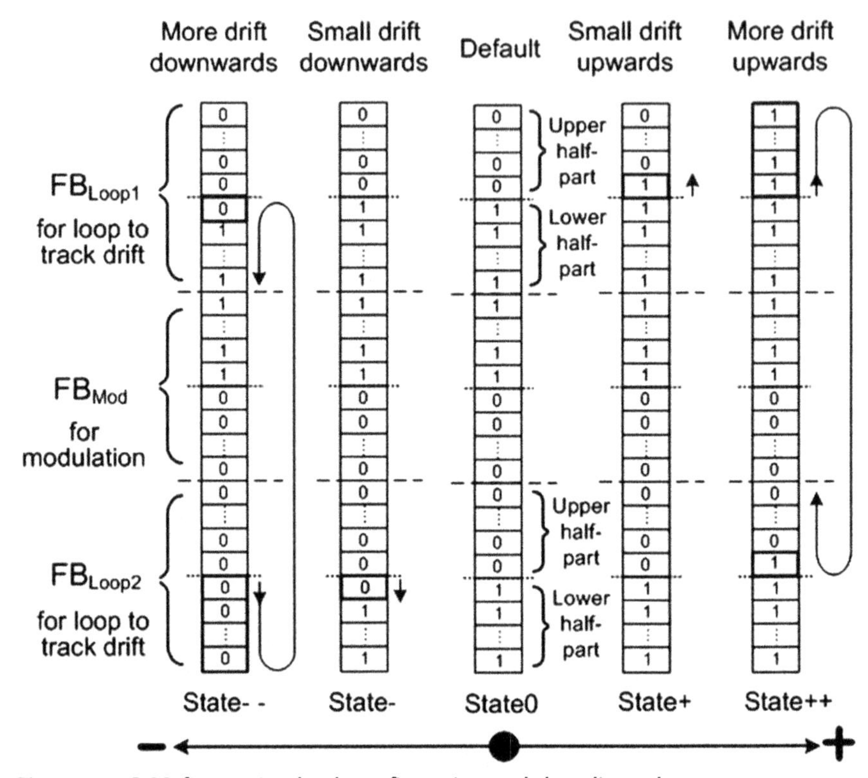

Figure 6.5 DCO fine-tuning bank configuration and decoding scheme.

1-MHz offset. This is significantly lower than the intrinsic DCO phase noise of −92 dBc/Hz at 1-MHz offset. The 60-GHz DCO drives the divider chain directly and is transformer-coupled via the FB structure to drive a three-stage neutralized PA. Thus, capacitive loading by the PA on the 60-GHz DCO core is attenuated by a factor of 10 via the weakly coupled transformer. Extensive EM simulations of the complete DCO tank together with interconnections to the divider and PA are performed with the EMX™ simulator [20] in order to verify the correct DCO tuning characteristic.

6.4 DIVIDER CHAIN DESIGN

A block diagram of the divide-by-32 chain is shown in Figure 6.6a. It consists of a 60-GHz LC-based ILFD, a high-speed CML divider with a maximum toggle frequency of 34 GHz, a divide-by-4 CML stage, and finally a

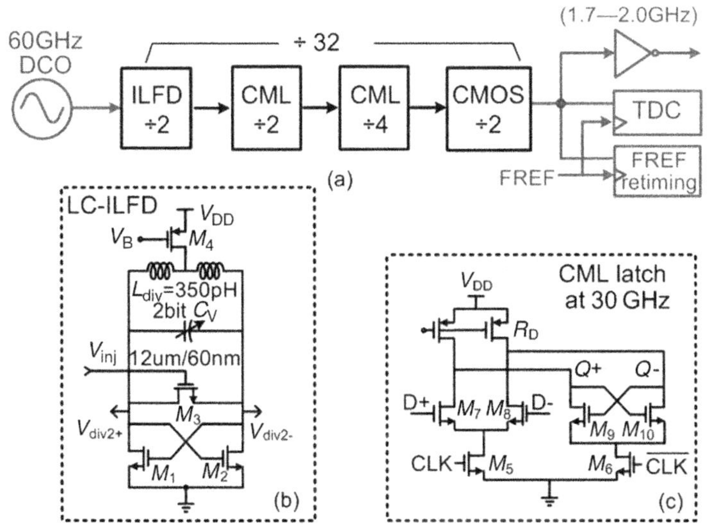

Figure 6.6 (a) Divide-by-32 chain in the 60-GHz ADPLL; (b) schematic of 60-GHz LC-ILFD; and (c) schematic of the 30-GHz CML divider.

CMOS divider. The signal at the divider chain output is approximately 2 GHz, and its phase is digitized via an incrementor plus a TDC for phase/frequency comparison with the frequency reference. As explained in Section 3.2, different divider topologies are used in each stage of the divider chain of this mm-wave frequency synthesizer to obtain a robust division over PVT with minimal power consumption and chip area.

6.4.1 Circuit Design

Detailed schematics for the first two divider stages are plotted in Figure 6.6b and c. These stages consume the lion's share of the power due to the operation in the 60- and 30-GHz bands, respectively. In the ILFD, a single-ended 60-GHz signal is injected directly into an LC resonator via M_3 ($W/L = 12$ µm/60 nm). A dummy cell is added to balance the capacitive loading of the ILFD on the differential outputs of the DCO. A simulated locking range of 9-GHz is realized by employing a 350-pH tank inductor and minimizing the total parasitic capacitance in the tank. The ILFD consumes 4.5 mA from a 1.2-V supply and delivers 500-mV single-ended peak voltage swing to the following CML divider. Switched-capacitor tuning of the ILFD free-running frequency (2 bits) extends the locking range (51–69 GHz), which is sufficient to cover PVT variations.

The high-speed CML latch employed in the 30-GHz prescaler is shown in Figure 6.6c. When the input clock (CLK) is high (M_5 ON), the input pair $M_{7,8}$ tracks (linearly amplifies) the input; when the clock is low (M_6 ON), the regenerative pair $M_{9,10}$ latches the state. Similar to CML buffers, a CML latch can operate with $2V_{THN}$ peak-to-peak voltage swings in the differential-mode (where $V_{THN} = 0.5$ V is the NMOS threshold voltage) [21]. Thus, the ILFD output is large enough to drive the CML latch directly. Unlike the latch used in the following divide-by-4 stage (as shown in Figure 3.12b), the tail current source is removed to increase the dc voltage drop across the load and the overdrive voltage of cross-coupled transistors $M_{9,10}$. Resistor R_D is provided by a PMOS transistor operating in triode to obtain a smaller voltage drop compared to a poly resistor of the same resistance value. A PMOS transistor is also preferred to an inductor load (as in Refs [22,23]), in order to place the latch close to the ILFD to form a compact layout. The tracking ($M_{7,8}$) and latch stages ($M_{9,10}$) are now optimized separately for correct operation at-speed. Cross-coupled pair $M_{9,10}$ (6-μm width FETs) provides the minimum required gain to hold the state (i.e., $g_{m9,10}R_D > 1$), while contributing only a small parasitic capacitance at nodes $Q+$ and $Q-$. The simulated maximum operating frequency in the worst case (i.e., SS process corner) is 34 GHz while consuming 8 mA from a 1.2-V supply. The CML divide-by-4 stage employs the circuit previously described in Figure 3.12. It operates from 8 GHz to 18 GHz and consumes 2.5 mA.

The last divide-by-2 stage in the prescaler (operating at 4 GHz) employs a CMOS divider with square-wave internal signals to achieve a much lower noise floor than the CML divider (e.g., −165 dBc/Hz). The schematic of the CMOS divider is similar to the one described in Figure 3.14. Two cross-coupled connections are used for quadrature generation. It consumes 1.5 mA while operating up to 5 GHz in the SS process corner (slow NMOS and slow PMOS). An inverter buffer with resistive feedback amplifies the CML divide-by-4 output to be able to drive the CMOS divider. The entire divider chain consumes 23 mA including interstage buffers, and delivers a rail-to-rail clock at approximately 2 GHz for the rest of the loop.

6.4.2 Simulated Performance

The divide-by-32 chain is simulated for various process corners: typical NMOS and typical PMSO (TT), fast NMOS and slow PMOS (FS), slow NMOS and fast PMOS (SF), fast NMOS and fast PMOS (FF), and slow

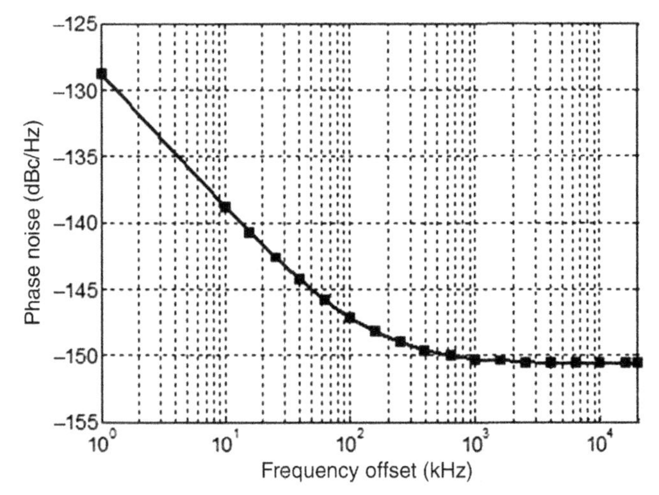

Figure 6.7 The simulated PN at divide-by-32 output (TT, 50°C).

NMOS and slow PMOS (SS) with the temperature variation from 0°C to 100°C, in the TSMC 65-nm CMOS technology. The 60-GHz band input with a single-ended peak-to-peak swing of 0.8 V is used in the simulation, corresponding to the worst case (i.e., SS corner) DCO output swing when loaded by the dividers and the PA. The divider-by-32 chain operates properly from 52 GHz to 66 GHz across all of the process corners and the entire temperature range. The simulated PN at the cascaded divider chain output is plotted in Figure 6.7. It has a noise floor of −150 dBc/Hz.

The simulated transient waveforms at the outputs of ILFD, high-speed CML divider, and the CMOS divider are shown in Figure 6.8. The 2-GHz square-wave output will be used to drive the high-speed CKV accumulator and the TDC as well. The input sensitivity curves of the entire divider chain are plotted in Figure 6.9 for the 2-bit switched-capacitor control. Each sub-band achieves 11-GHz locking range with maximum input power of +5 dBm. The entire divider chain can operate from 51 GHz to 67.5 GHz with 2-bit switched-capacitor bank control, while consuming 23 mA from a 1.2-V supply.

When used in the ADPLL, open-loop calibration is first performed to obtain the tuning range of the 60-GHz DCO and the self-resonant frequency of the 60-GHz ILFD. Then the 2-bit switched-capacitor bank control of the ILFD is manually configured to ensure that the locking range of the ILFD overlaps with the DCO tuning range.

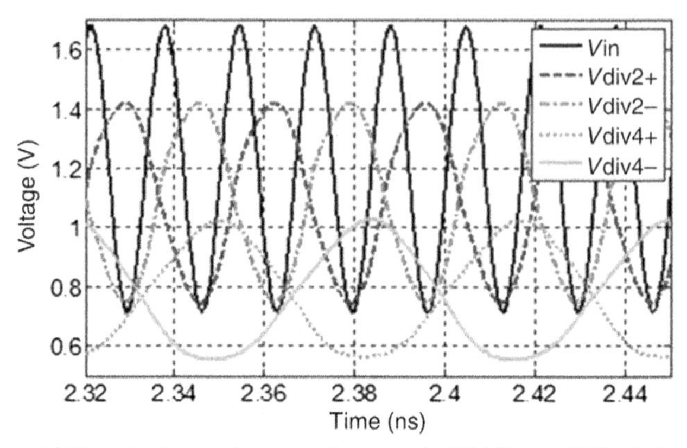

(a) Transient waveforms at the output of ILFD and CML divier.

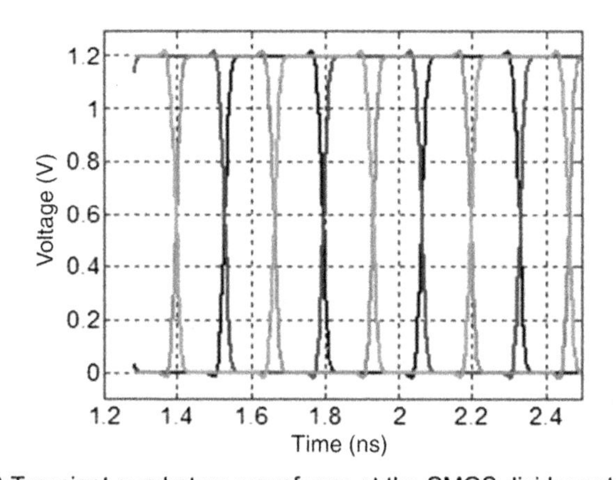

(b) Transient quadrature waveforms at the CMOS divider output.

Figure 6.8 Simulated transient waveforms along the divider-by-32 chain (TT, 50°C).

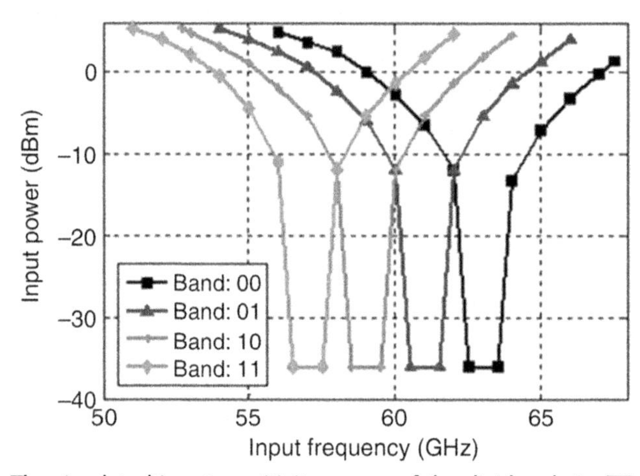

Figure 6.9 The simulated input sensitivity curves of the divider chain (TT, 50°C).

6.5 TDC DESIGN AND CALIBRATION

The TDC is a key block in the ADPLL architecture and replaces the phase/frequency detector and charge-pump of a conventional PLL synthesizer. It is used to generate the variable phase (PHV) as shown in Figure 6.10. The resolution of the TDC must be a small fraction of the DCO period to guarantee an acceptable level of in-band phase noise. For the intended FMCW radar applications, the required TDC resolution is 12 ps, as listed in Table 6.2. A TDC based on a pseudo-differential delay chain was first reported in Ref. [24]. That TDC structure is quite simple and appears sufficient for the 12 ps resolution target. The TDC circuit and calibration techniques are elaborated in this section. Other TDC circuits, which achieve finer time resolution than with the simple delay-line-based structure, will be discussed at the end of this section.

6.5.1 TDC Core Architecture

Figure 6.11 depicts the TDC core architecture. It is pseudo differential to avoid any mismatch between rising and falling edge transitions due to

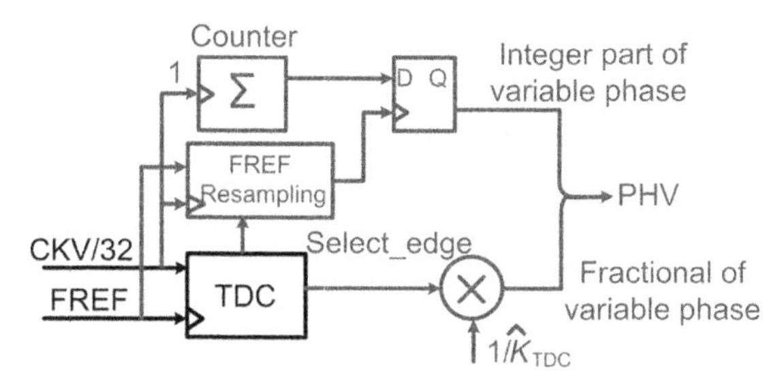

Figure 6.10 TDC in the ADPLL for variable phase (PHV) generation.

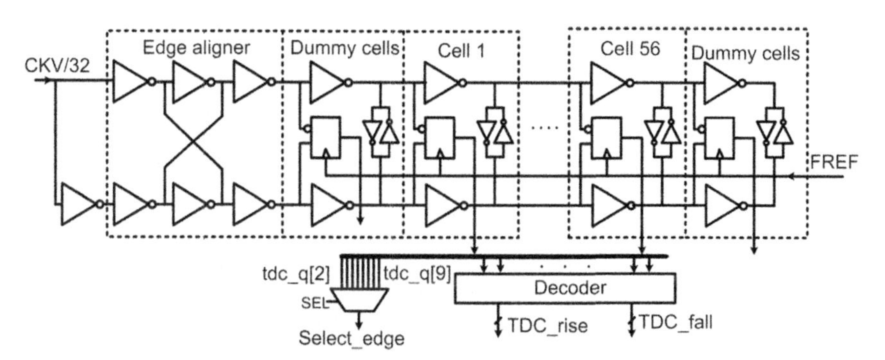

Figure 6.11 TDC core architecture.

differing strengths of the NMOS and PMOS transistors. To compensate for any phase difference in the differential (or rather, complementary) high-speed clocks (CKV/32) due to routing mismatches, they are generated locally from a single-ended input and then passed through a complementary string of 56 inverters.

Four dummy cells are placed at both the beginning and the end of the TDC delay chain to improve the mismatch of the unit delay cells. The edge aligner shown in Figure 6.11 generates differential CKV/32 signals locally. The delayed-clock replica vector is sampled by FREF using an array of 50 sense-amplifier-based flip-flops (SAFFs) that are adapted from Ref. [25]. The 56-bit TDC output forms a pseudo thermometer code, which is then converted to binary (TDC_rise and TDC_fall) using a simple digital priority decoder [24]. The decoder searches for the first one-to-zero transition that measures the CKV/32-to-FREF delay in units of an inverter delay to obtain TDC_rise. The number of inverters is set to cover one T_v, which is the period of CKV/32. The largest T_v corresponds to the lowest DCO output frequency and equals 560 ps (i.e., 32/57 GHz). The select_edge signal shown in Figure 6.11 is derived from the TDC delay chain to choose the path farther away from the metastable region during FREF retiming, in which FREF is oversampled by both rising and falling edges of CKV/32, simultaneously.

The absolute phase of CKV/32 is more useful than the instantaneous frequency between its edges. Also, the CKV/32 edge locations versus the reference edges are quite predictable, so significant power savings are made by turning them OFF for more than 90% of the time between the reference edges, as shown in Figure 6.12.

6.5.2 TDC Unit Cell Design

The schematic of the TDC unit cell is shown in Figure 6.13, which consists of two unit-inverter delay cells, a back-to-back inverter pair connected in between, and a sense-amplifier flip-flop. In the unit-inverter delay cell, both NMOS and PMOS use minimum-length transistors with

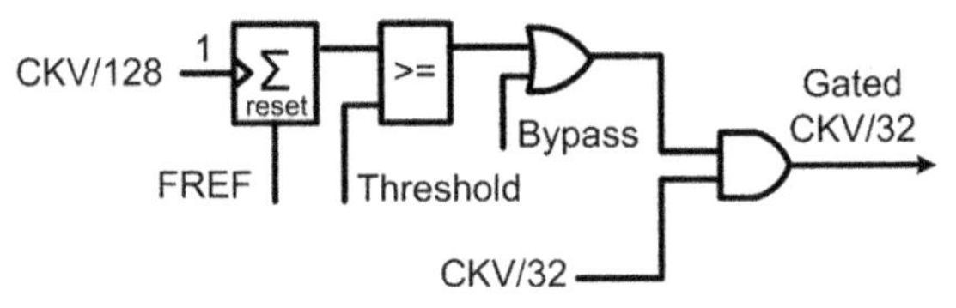

Figure 6.12 TDC system with power saving through clock gating.

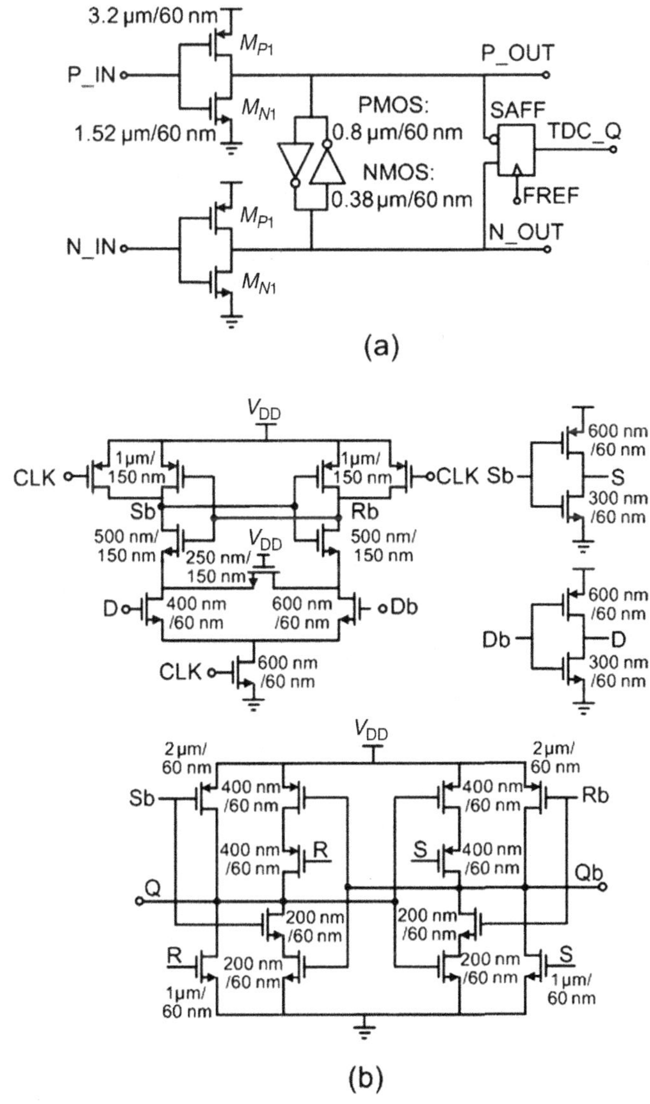

Figure 6.13 (a) Schematic of TDC unit cell and (b) schematic of SAFF.

a width ratio of 1:2.1 to achieve substantially equal rise and fall times. The width of both NMOS and PMOS are sized up proportionally to the values shown in Figure 6.13a in order to obtain close to the intrinsic delay time (Δt_{inv}) in the 65-nm CMOS process ($\sim$12 ps). Increasing the inverter size further will have marginal improvement on the time resolution, but consumes more power. The small back-to-back inverter pair connected between the two delay chains helps to further improve the even–odd mismatch (i.e., delay difference between the two adjacent unit delay cells in the TDC chain) due to the unequal rise and fall times.

The schematic of the SAFF is depicted in Figure 6.13b. The SAFF is symmetrical about the vertical axis and provides identical resolution of metastability at the rising or falling edge of their input data. Capacitive loading of the data input is only one NMOS gate, and great effort is spent during the layout so that the interconnect parasitics are minimized. The metastability window is very small: within a resolution time of 1 ns, the metastability resolution window is much less than 1 ps, which is much smaller than the inverter delay. Further allowing the resolution time to increase will bring a rapid exponential decrease of the resolution window. This easily ensures no "bubbles" (i.e., thermometer code is not corrupted due to metastability) in the TDC code and makes inverter delay time the only factor that limits the TDC resolution.

6.5.3 TDC Calibration

The TDC is also self-calibrating for the inverter delay variations with respect to PVT during the regular operation, as shown in Figure 6.14. The absolute difference between the measured rising (TDC_rise in Figure 6.11) and falling (TDC_fall) edge delays of CKV/32 to the closest FREF rising edge is the half-period of HCLK in terms of number of inverters (i.e., one-half of $T_v/\Delta t_{\text{inv}}$). An accurate estimate of $T_v/\Delta t_{\text{inv}}$ is obtained through averaging over 2^{10} samples, with an error below 1%. Its inverse is used for the fixed-point period normalization multiplier with the 19 fractional bits (W_F). This value divided by the CKV/32 frequency ($1/T_v$) is the inverter delay in units of second. The normalized TDC output produces the fractional part of the variable phase in Figure 6.10, which is used in the phase error detection.

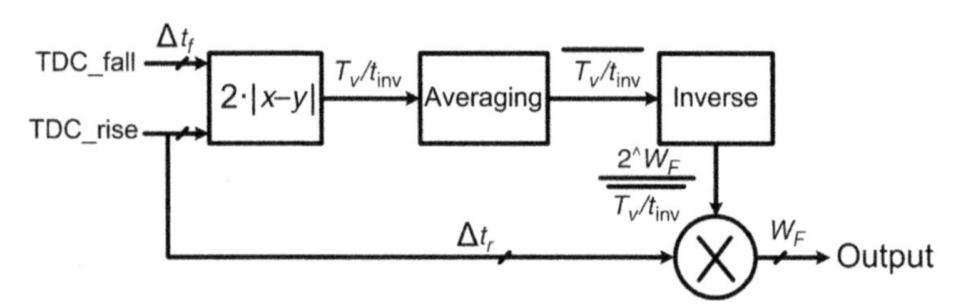

Figure 6.14 TDC gain calibration.

6.5.4 Simulated Performance

The compact layout of the TDC is critical to achieving good linearity and fine time resolution. It is revealed in Figure 6.15. The TDC unit cell includes two main inverters, one SAFF and two small back-to-back regenerative inverters. The two pseudo-differential delay chains are symmetrical and the interconnections between adjacent unit cells are minimized to reduce capacitive loading and the time delay. The entire 64 cells (56 effective cells and eight dummy cells) are placed in one row. The high-frequency clock HCLK is fed into the TDC from left to the right, while the low-frequency reference clock FREF travels from right to the left. The entire TDC is 175 μm × 30 μm. The simulated TDC time resolution t_{res} for PVT variations is shown in Figure 6.16, which varies from 8.648 ps (the best case: FF corner, $V_{DD} = 1.3$ V, and $T = 0°$C) to

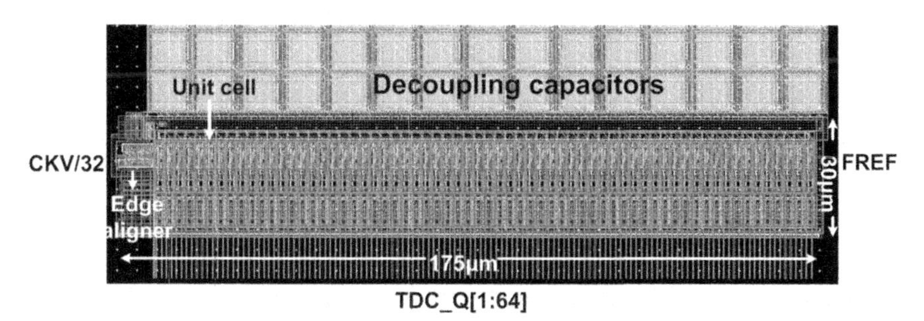

Figure 6.15 Layout of the TDC.

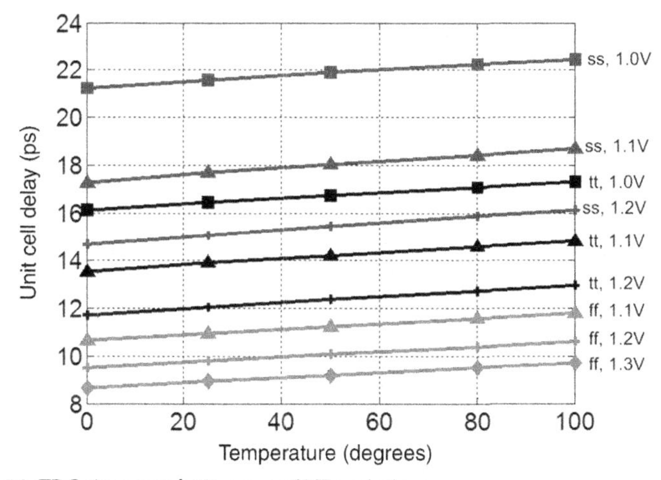

Figure 6.16 TDC time resolution over PVT variations.

22.43 ps (the worst case: SS corner, $V_{DD} = 1$ V, and $T = 100°C$). The Monte-Carlo device mismatch simulation results of the t_{res} are plotted in Figure 6.17, for both even and odd cells (at TT corner, $V_{DD} = 1.2$ V, and $T = 27°C$). The mean value of t_{res} is 12.1 ps for the entire TDC delay chain with a standard deviation of 0.9 ps. The t_{res} for even and odd cells are 11.94 and 12.25 ps, respectively, due to the small asymmetry in the rising and falling edges of the unit inverter cell with layout parasitics. The simulated metastability window of the SAFF is shown in Figure 6.18. It is

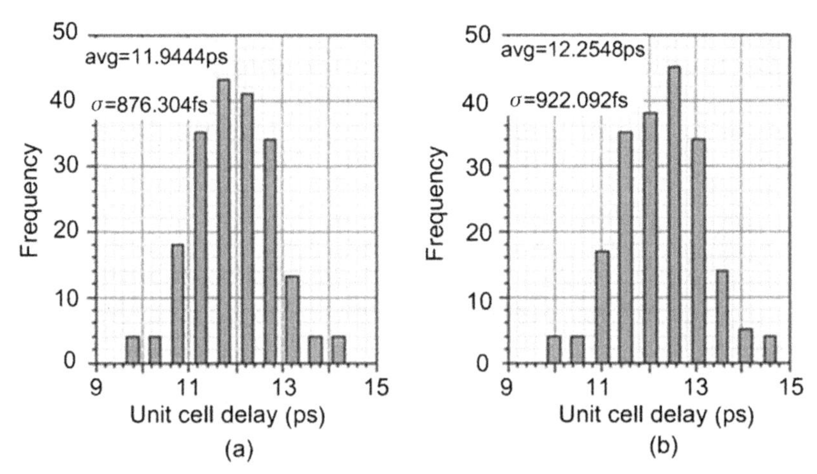

Figure 6.17 Monte-Carlo simulation results of TDC unit cell delay: (a) even cell and (b) odd cell cells (nominal case (TT), $V_{DD} = 1.2$ V, and $T = 27°C$).

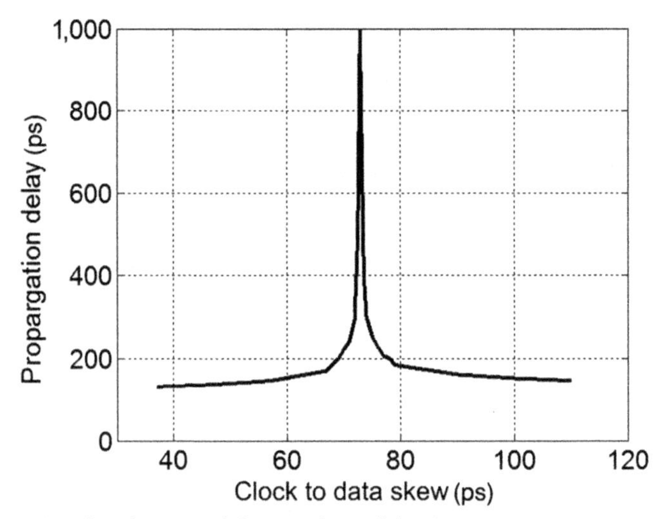

Figure 6.18 Simulated metastability window of the SAFF.

symmetrical about the vertical axis in the middle of the plot. The TDC consumes 5.5 mA from a 1.2-V supply when clock gating is not enabled and the current consumption reduces to only 1.5 mA when the clock gating shown in Figure 6.12 is enabled.

To overcome the limit set by the minimum delay available in the given technology, several improved TDC architectures have been reported in the literature. Linear Vernier TDCs quantize time differences exploiting the cumulative delay difference of two lines based on elements whose delay is greater than the target resolution [26]. This technique breaks the trade-off between minimum stage delay and TDC resolution, but the number of delay elements grows exponentially with the number of bits limiting the advantages when a fine resolution is required. The large number of delay stages increases the TDC integral nonlinearity (INL). Two-dimensional Vernier TDC in Ref. [27] exploits all possible time differences between the taps of the two delay lines in order to decrease the number of stages of the delay lines and, thus, to reduce the complexity and the power consumption of the structure. However, the TDC transfer function is still nonlinear due to the mismatches between delay elements and so a calibration would normally be required. Alternatively, gated ring oscillators (GROs) [28,29] can serve as TDCs with noise-shaping characteristics. GRO-based TDC provides fine time resolution with better linearity as the mismatches between the delay elements are also shaped to the higher-frequency offset via the intrinsic $\Sigma\Delta$ operation. However, due to leakage, the GRO internal state could vary during the off-state and so it translates to an elevated noise floor. This issue is known as GRO timing skew and should be taken care of in high-performance TDC designs. In addition, coarse-fine TDC [30,31] and interpolative approach [32] are also commonly found in the literature to improve the TDC resolution while maintaining the required dynamic range.

6.6 REFERENCE SLICER DESIGN

The reference slicer (or buffer) generates an on-chip frequency reference (FREF) for the PLL from an off-chip crystal oscillator, as shown in Figure 6.19. The thermal and flicker noise of the slicer contributes to the in-band PN of the PLL, and thus must be kept well below the contribution due to the TDC quantization noise in order not to be the dominant noise factor.

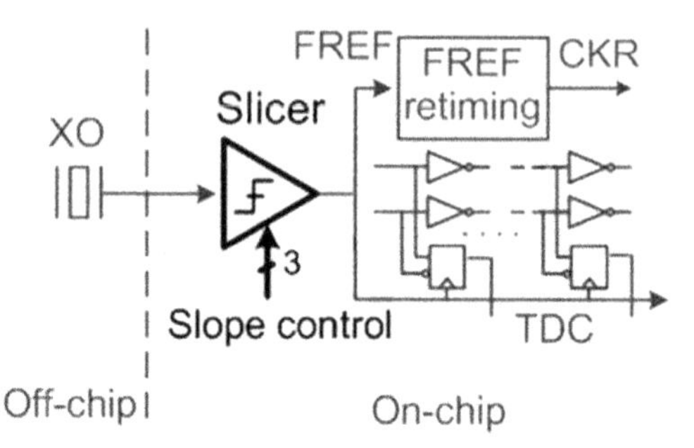

Figure 6.19 Reference slicer (or buffer) in the ADPLL system.

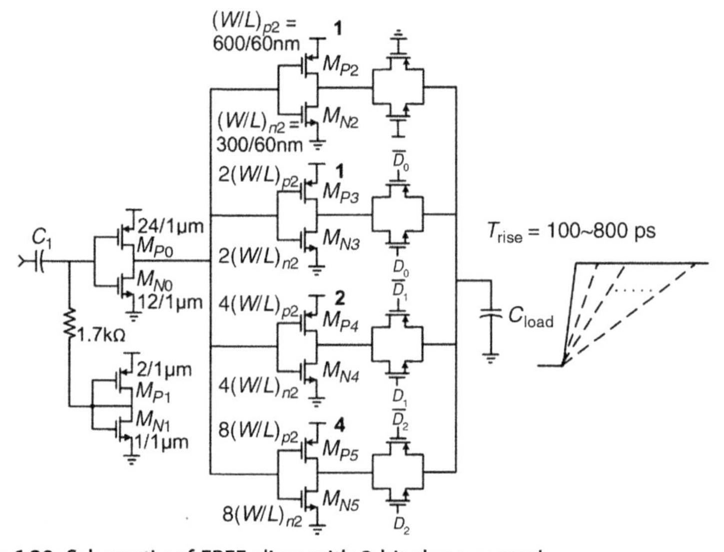

Figure 6.20 Schematic of FREF slicer with 3-bit slope control.

6.6.1 Circuit Design

The schematic of the slicer is shown in Figure 6.20. The external reference signal of 200 mV peak swing is AC-coupled and amplified to a rail-to-rail swing by the inverter amplifier consisting of M_{P0} and M_{N0}. The transistor channel length is sized to 1 μm to suppress flicker noise for both M_{P0} and M_{N0}. A small-sized inverter replica (M_{P1} and M_{P2}) is configured as a unity-gain buffer to provide gate bias for the main amplifier. The output of the first-stage amplifier drives four independent paths, each consisting of a digital inverter cascaded with a transmission gate. The overall driving capability at the slicer output is binary controlled (3 bits) by transmission gates in each path in order to exert control on the rising/falling slopes of FREF. By slowing down the FREF transition time from 100 to 800 ps, higher-order harmonics generated during transitions are attenuated and the reference spur can be attenuated by at least 2 dB in the measured PLL output spectrum. No degradation of the PLL in-band phase noise was detected, since it is dominated by the TDC quantization noise. To achieve a (low) phase noise of −135 dBc/Hz at 10-kHz offset, any on-chip interference at the slicer is minimized by placing the slicer input close to the bondpad. The rail-to-rail swing of the slicer output is then fed to the TDC via an on-chip microstrip line of 100-μm length, which further slows down the rising edge of FREF as it is distributed on-chip.

6.6.2 Simulated Performance

The simulated transient waveforms at the slicer input and output are plotted in Figure 6.21. An external crystal oscillator of 40 MHz provides a sinusoidal input signal with approximately 400-mV peak-to-peak swing to the slicer. The slicer output is a rail-to-tail swing signal proving reference clock FREF to the PLL. The simulated PN performance at the slicer output is −138.78 dBc/Hz at 10-kHz offset from the 40-MHz carrier for the fastest FREF slope (plotted in Figure 6.22). The PN degrades to −137 dBc/Hz for the slowest FREF rising edge. The PN of the reference clock is much lower than the PN contribution due to the TDC

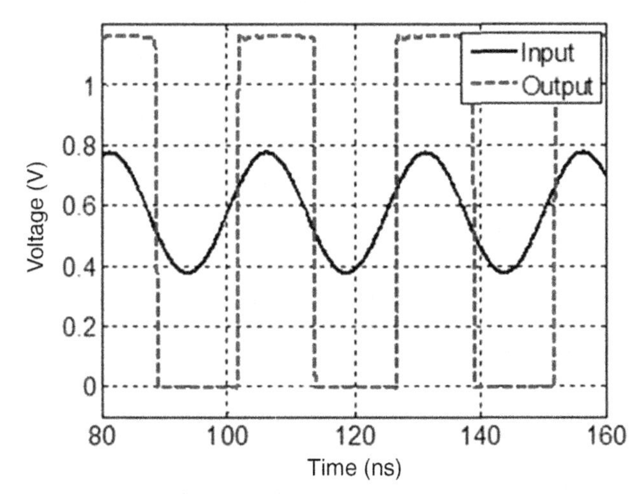

Figure 6.21 Transient waveforms at slicer input and output.

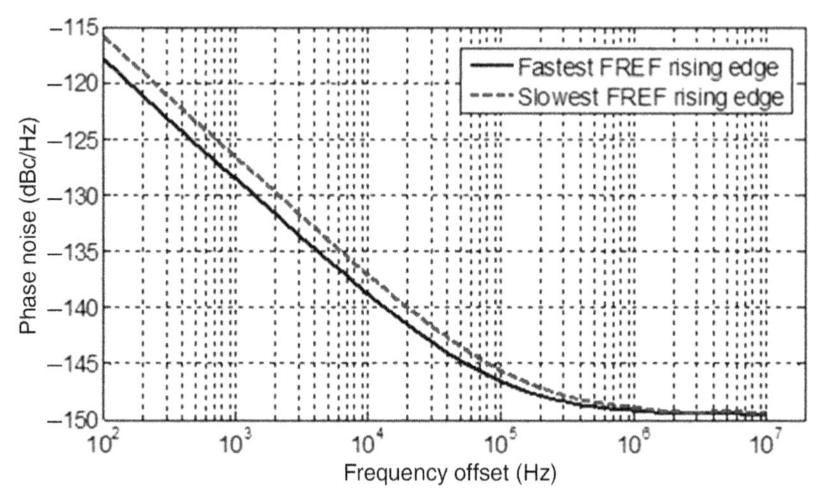

Figure 6.22 Simulated phase noise at slicer output.

quantization noise, and thus has negligible contribution to the in-band phase noise of the PLL.

6.7 PHASE ERROR GENERATION AND GLITCH REMOVAL

As explained in Section 6.2, the 60-GHz ADPLL operates in the digitally synchronous fixed-point phase domain. The reference phase is obtained by accumulating FCW at each rising edge of the retimed frequency reference. For the variable phase signal (Figure 6.23), its integer part is determined by counting the number of rising clock transitions of the variable clock (CKV/32), while the fractional part is obtained by the TDC described in Section 6.5. Samples of the variable phase signal are subtracted from the reference phase in a synchronous arithmetic phase detector to determine the phase error. As described in Ref. [33], the sampling moments of the accumulator value are not the same as those of the TDC. The FREF clock provides triggering moments which sample both the counter and TDC outputs. These different sampling instances could have a substantial timing misalignment τ, indicated in Figure 6.23a, and thus cause glitches in the phase error when the counter and TDC outputs are combined (see Figure 6.23b).

Instead of correcting these glitches [33], a simple glitch removal method is proposed, as shown in Figure 6.23c. Digital logic first detects a glitch by comparing the current phase error to that in the previous clock

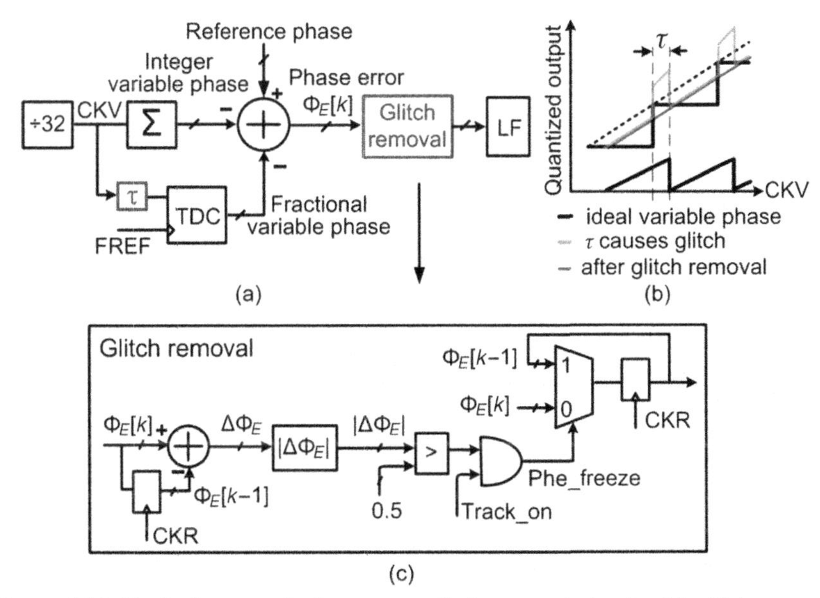

Figure 6.23 Block diagram of phase error glitch removal circuits: (a) glitch removal circuit in a digital PLL; (b) quantized phase error before and after glitch removal; and (c) implementation detail of glitch removal circuits.

cycle. If the difference is larger than a certain threshold (e.g., 0.5), the input is assumed to contain a glitch. The phase error is then frozen for this clock cycle by simply disregarding the current phase error. This method introduces a constant phase offset (Figure 6.23b) in the glitch-free phase error output, which is harmless to the ADPLL operation. In addition, the same logic can be re-used as a lock indicator, or to generate a clock quality monitoring signal for design-for-test purposes, which will be elaborated on in Chapter 8.

The glitch-free phase error is then filtered by a dynamically reconfigurable loop filter to tune the DCO to the desired frequency and phase. The reconfigurable loop filter implementation and the gear-shifting scheme are similar to the digital LF design described in Section 4.3.

6.8 TOP-LEVEL FLOOR PLAN CONSIDERATIONS FOR MM-WAVE ADPLL

Parasitic spur suppression and injection pulling should be carefully addressed in the top-level floor plan of the ADPLL-based transmitter IC. Higher harmonics generated by the digital circuitry fall into the vicinity of the DCO LC-tank resonant frequency. The coupling mechanism could be magnetic (DCO inductor, bondwires), capacitive electric (long parallel wires), through the substrate, through common ground/supply connections, etc. Injection-pulling in the ADPLL is reduced by the FREF retiming mechanism. The digital circuitry and memory interfaces are clocked synchronously with the RF oscillator (a retimed FREF), thus mitigating their aggressor effects despite the fractional frequency relationship between the aggressor's harmonic and the RF carrier. Another major coupling mechanism in the ADPLL-based transmitter [34] is as follows: the FM/PM modulated RF clock in the feedback path to the TDC (i.e., an aggressor) gets parasitically coupled into the slicer input (i.e., a victim) of the FREF crystal oscillator. By means of subharmonic modulation of the sinusoidal FREF oscillator waveform, jitter is created on the FREF digital clock. The jitter energy passes through the low-pass loop filter and modulates the DCO. This coupling mechanism creates significant distortion of the modulated RF waveform at the near integer-N channels (i.e., when the instantaneous FCW is very close to an integer). Fortunately, the output frequency is always ramping up and down linearly in an FMCW radar application, and thus the FCW will not stay at the near integer-N channel for more than a few cycles.

The floor plan of the 60-GHz FMCW transmitter is shown in Figure 6.24. It contains five different power domains and three ground domains. Each critical analog building block (i.e., DCO, TDC, and reference slicer) has a P + guard-ring around it and local supply decoupling. The entire digital part is located on the right-hand side of the layout,

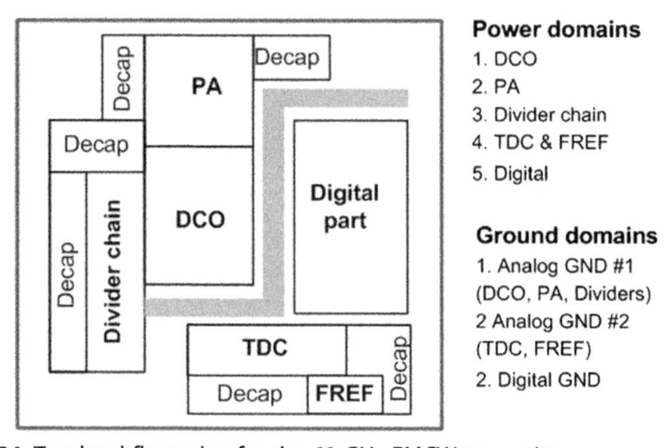

Figure 6.24 Top-level floor plan for the 60-GHz FMCW transmitter.

which is also enclosed by a P+ guard-ring to isolate it from the analog part. Grounds for the digital and analog parts are separated on chip, and are joined together off-chip on a printed circuit test board. An isolation region between the digital and analog portions (gray region in Figure 6.24) is inserted, in which the substrate resistivity is 10 times higher than the active region to improve the isolation.

6.9 EXPERIMENT RESULTS FOR 60-GHZ ADPLL

The 60-GHz ADPLL described above is transformer-coupled to a three-stage power amplifier to form a FMCW transmitter. The prototype was fabricated in TSMC 65-nm CMOS with one poly and seven metal layers [35]. The die micrograph is shown in Figure 6.25. The 2.2-mm^2 total die area includes bondpads, PA, SRAM, and other digital circuitry for debugging. The ADPLL core occupies 0.5 mm^2 and the total on-chip SRAM is 6×2^{13}-bit. The DC pads for the analog sections are located on the left-hand side while the digital IO and DC pads are on the right hand-side. The 60-GHz PA output and the other RF/analog test outputs are on the top and bottom sides, respectively. The ADPLL chip consumes 40 mA: 11 mA by the DCO, 23 mA in the frequency prescaler, 6 mA by the TDC and the digital logic/memory, while the PA dissipates 34 mA,

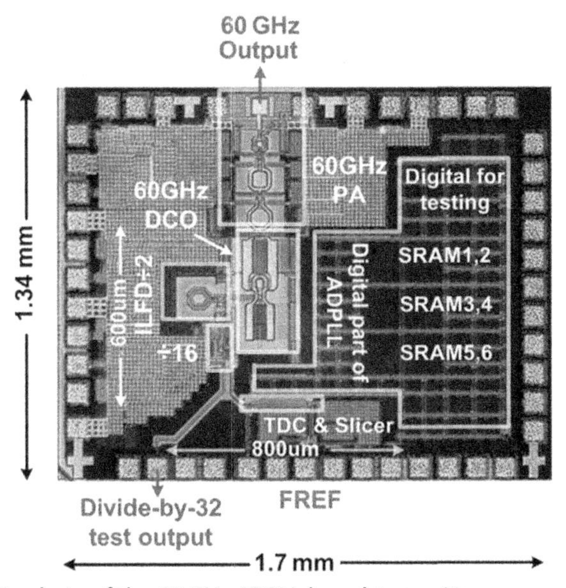

Figure 6.25 Die photo of the 60-GHz ADPLL-based transmitter.

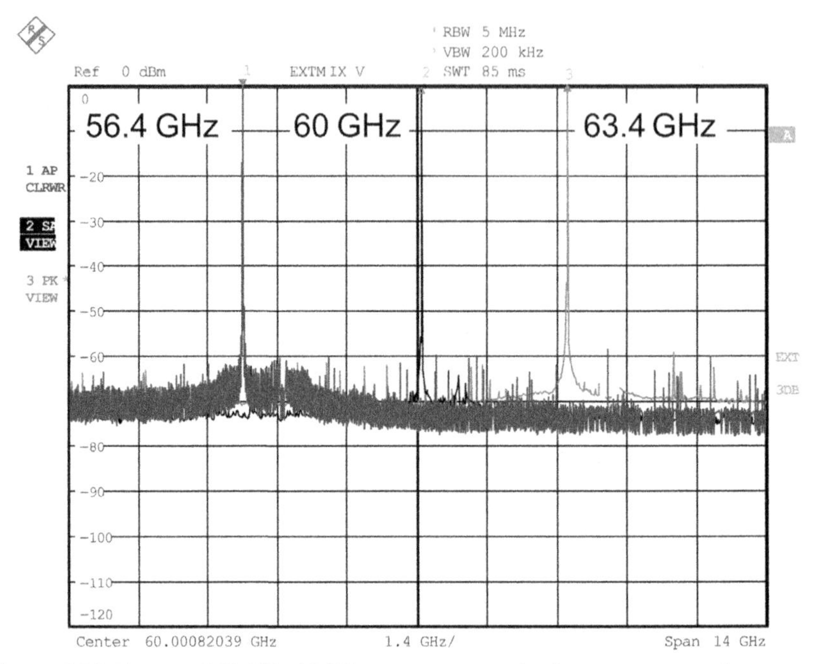

Figure 6.26 Measured 60-GHz ADPLL spectra across the frequency operating range.

all from a single 1.2-V supply. The testing setup and the design-for-test issues will be discussed in Chapter 8. The wideband linear modulation results will be revealed in Chapter 7, following the discussion on time-domain calibration techniques. In this section the key measurement results of the 60-GHz ADPLL are examined and compared to state-of-the-art analog PLLs.

The 60-GHz fractional-N ADPLL can generate arbitrary frequencies from 56.4 GHz to 63.4 GHz. The spectra of the locked synthesizer output across the 60-GHz band are all plotted in Figure 6.26, in which no spurious tones above −60 dBc are observed in a 14-GHz wide span. The actual spurious tones at the 60-GHz synthesizer output are much lower than −60 dBc. They are due to spurious tones generated by the subharmonic mixer (1/8th) used in the setup. This is confirmed by the measured spurious free wide span spectra at the divide-by-32 output in Figure 6.27.

The measured spectrum of the mm-wave output when the ADPLL is locked at 60.02 GHz is plotted in Figure 6.28a. A very low reference spur level of −74 dBc is observed, with no other significant spurs detectable.

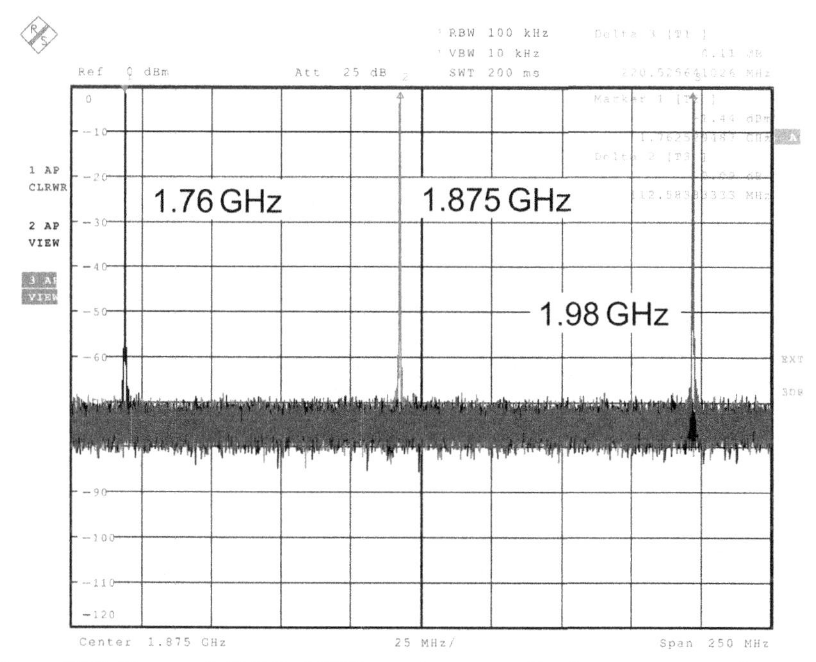

Figure 6.27 Measured 60-GHz ADPLL spectra across the frequency operating range at the divide-by-32 output.

The measured worst-case reference spur is −72.4 dBc across the 7-GHz locking range. The PLL output spectrum is also measured for a larger frequency span (e.g., 1 GHz) and no other significant spurs are observed. The out-of-band fractional spurs are filtered out heavily by the type II, fourth-order IIR loop filter. For some FCW (e.g., near integer-N channels), the fractional spurs fall in-band (cannot be filtered by the loop filter) but are always less than −60 dBc. The spurious tones due to the TDC nonlinearity and finite TDC resolution at near integer-N channels are not specifically addressed in this design, since the synthesizer output will always be modulated by a triangular wave. Thus, the TDC input will only remain at the near integer-N channel for a short period of time (e.g., a few FREF cycles).

When the phase error glitch removal logic is disabled, the skirt level of the PLL output spectrum increases by 2 dB. The close-in spectrum of the 60-GHz carrier is shown in Figure 6.28b, which reveals that the PN at 1-MHz frequency offset is −89 dBc/Hz ($-42.05 - 10\log_{10}[50\ \text{kHz}]$). Figure 6.29 compares the PLL output spectra while in free-running and in lock. It is clear that a "shoulder" is formed in the locked spectrum by

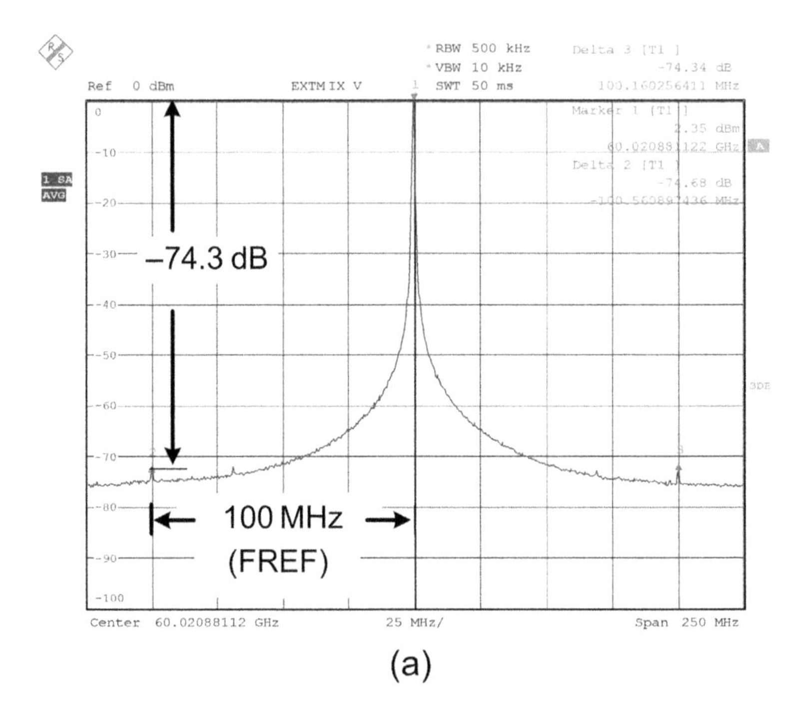

(a)

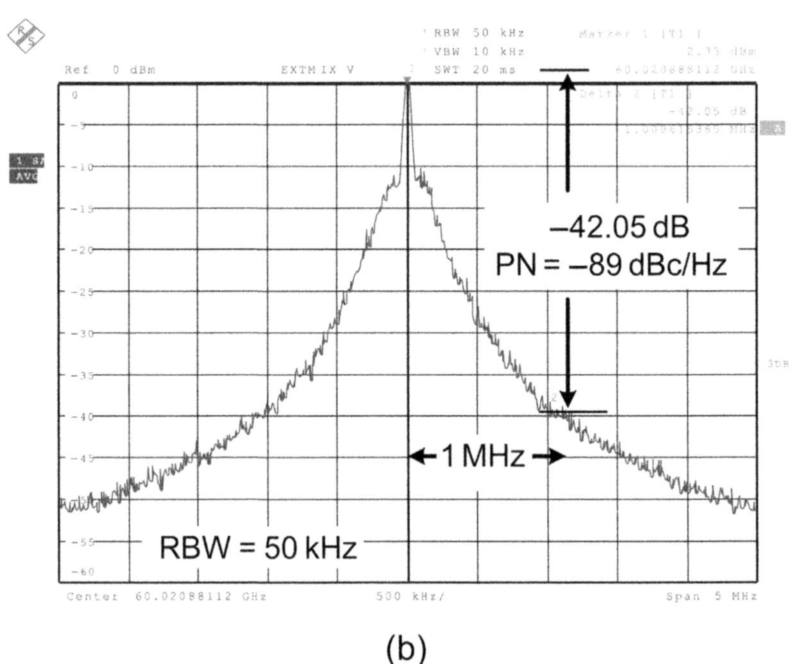

(b)

Figure 6.28 Measured 60-GHz ADPLL spectrum.

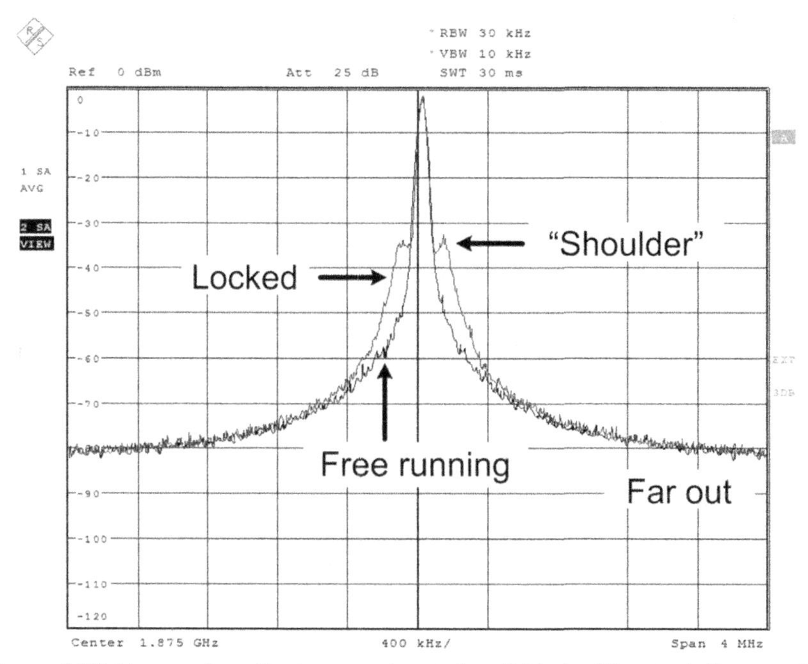

Figure 6.29 Measured synthesizer spectra at the divide-by-32 output (free-running and locked).

the TDC noise in combination with the DCO noise, as set by the PLL bandwidth. The two spectra are very close at the far out region, which indicates that the loop operation does not contaminate the DCO PN.

To simplify the measurement, the PN of the ADPLL is measured at the divide-by-32 (CKV/32) test output (see Figure 6.30 for the PN measured at various loop bandwidths). Thus, 30.1 dB (i.e., $20\log_{10}[32]$) should be added to refer the PN to the mm-wave (i.e., 60 GHz) output. For a loop bandwidth of 300 kHz, the measured PN at the divide-by-32 output is −118 dBc/Hz at 1-MHz offset, which agrees well with the PN obtained at the PA output shown in Figure 6.28b. Compared to the free running DCO PN of −92 dBc/Hz at 1-MHz offset, the synthesizer PN degrades by 4 dB under this particular LF configuration. The integrated PN measured from 10 kHz to 10 MHz at the divide-by-32 output is −45.9 dBc (see Figure 6.30) for the 300 kHz PLL loop bandwidth. This corresponds to an integrated PN of −15.8 dBc ($= -45.9 + 20\log 32$) at 61.87-GHz output. The rms jitter is 590.2 fs. The measured integrated PN at the PLL output varies by 1.5 dB for the same LF configuration over the 7-GHz locking range.

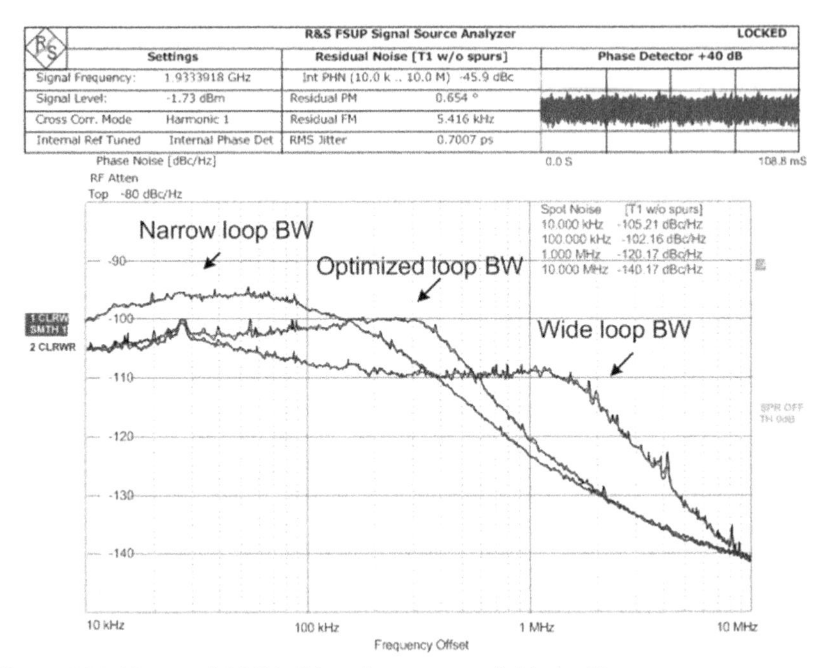

Figure 6.30 Measured ADPLL PN performance at divide-by-32 output.

The TDC is self-calibrating for the PVT inverter delay variations during normal operation of the ADPLL as described in Section 6.5.3. The calibration result indicates the average inverter delay is 12.2 ps at room temperature, which corresponds to a theoretical in-band PN of −80 dBc/Hz for a 60-GHz carrier. From Figure 6.30, the measured in-band PN is −78 dBc/Hz at 60-GHz band for a wide loop BW (~1.5 MHz) and thus is dominant by the TDC quantization noise (−80 dBc/Hz). The TDC consumes 6 mA and its current consumption is reduced to only 1.5 mA when the power management unit in Figure 6.12 is active.

Aside from the above spectral purity, the lock-in time is another key metric for the synthesizer. To simultaneously achieve fast locking and excellent PN after settling, the loop bandwidth is dynamically controlled via a gearshift technique described in Section 4.3. The loop operates in type I with a wide bandwidth of 1.5 MHz during the frequency acquisition. It is then switched hit-lessly to a type II, fourth-order IIR filter with 300-kHz bandwidth only when it enters tracking mode. The measured lock-in time is 3 μs for a frequency step up to approximately 10% of the carrier frequency, as demonstrated in Figure 6.31.

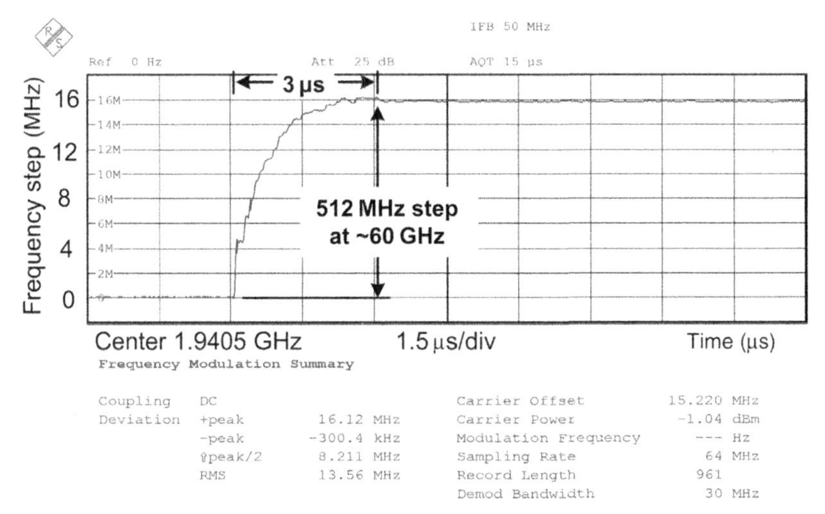

Figure 6.31 Lock time of the 60-GHz ADPLL measured at divide-by-32 output for a frequency step of 512 MHz.

As the first ADPLL reported at 60 GHz, the performance of the prototype is compared to leading 60-GHz analog PLLs [36–38], as well as ADPLLs operating at 40 GHz [39] in Table 6.3. The ADPLL example discussed in this chapter features fractional-N frequency precision, exhibits good in-band and out-of-band PN performance (−15.8 dBc integrated PN at 60 GHz), fast locking (within 3 µs), and an ultra-low reference spur (−74 dBc) when compared to these analog examples from the recent literature. The ADPLL also requires small chip area (core area of 0.4 mm^2) and power consumption (40 mA from 1.2 V).

6.10 SUMMARY

A 60-GHz fractional-N ADPLL implemented in 65-nm bulk CMOS for the FMCW transmitter applications is employed as a design example to explain the mm-wave ADPLL architecture and circuit design techniques. Compared to the conventional charge-pump PLL, the all-digital architecture features extensive reconfigurability and allows auto-calibration (more digital calibration techniques will be discussed in Chapter 7). The mm-wave DCO exploits distributed switched-metal capacitors for frequency tuning and a transformer to attenuate the frequency tuning step. It oscillates from 56.4 GHz to 63.4 GHz with a fine frequency resolution of approximately 1 MHz. The divide-by-32 chain can operate from 51 GHz to

Table 6.3 Performance comparison of the 60-GHz ADPLL with state-of-the-art 60-GHz analog PLLs and 40-GHz ADPLL

		60-GHz ADPLL	ISSCC'09 [36]	JSSC'11 [37]	ISSCC'13 [38]	TCAS-II'11 [39]
Center frequency		60 GHz	61.5 GHz	60.5 GHz	63.1 GHz	40 GHz
Architecture		TDC-ADPLL	Charge-pump PLL	20 GHz analog PLLx3	Charge-pump PLL	BB-ADPLL
Type		Fractional-N	Integer-N	Integer-N	Integer-N	Integer-N
Modulation		Yes (2-point)	No	No	No	No
CMOS		65 nm	45 nm	65 nm	65 nm	90 nm
Reference frequency		100 MHz	100 MHz	36 MHz	135 MHz	156.25 MHz
PN	$\Delta f = 10$ kHz	−75 dBc/Hz	−70 dBc/Hz	−60 dBc/Hz	−80 dBc/Hz	−75 dBc/Hz
	$\Delta f = 1$ MHz	−90 dBc/Hz	−75 dBc/Hz	−95 dBc/Hz	−91 dBc/Hz	−83.87 dBc/Hz
RMS jitter		590.2 fs	NA	NA	238.4 fs	300.87 fs
FREE spur		−74 dBc	−42 dBc	−67 ~ −58 dBc at 20 GHz	<−45 dBc	−48 dBc
Frequency range		7 GHz (11.6%)	9 GHz hi/lo, 2 VCOs, (14.6%)[a]	5 GHz (8.3%)	10.4 GHz (16.5%)	3 GHz (7.5%)
Locking time		3 μs	NA	NA	NA	NA
Supply		1.2 V	1.1 V	1.2 V	1.2 V	1.2 V
Power consumption		48 mW	78 mW[b]	80 mW[b]	24 mW[b]	46 mW
Core area		0.48 mm^2	0.82 mm^2 (with pads)	1.68 + 0.8 mm^2 (2 dice, with pads)	0.192 mm^2	0.3 mm^2

[a]PLL locking range is 2.4 GHz, limited by injection locking divider's locking range.
[b]Provides quadrature outputs at 60 GHz.

69 GHz, providing sufficient margin for PVT variations. The divider chain consists of a 60-GHz ILFD, a 30-GHz CML divide-by-2 stage, another CML divide-by-4 stage, and a CMOS divider, which consumes 23 mA in total (including inter-stage buffers). The delay-chain TDC achieved a measured time resolution of 12 ps at 1.2-V supply. The low-jitter slicer with 3-bit slope control minimizes reference spurs. The dynamically controlled LF supports multiple gearshift events, which can reduce or enlarge the PLL loop bandwidth smoothly without any phase error perturbations. The LF parameters are optimized for the lock-in process and the phase noise performance when the loop is settled, respectively. The glitch removal circuit eliminates the large phase jump due to the misalignment between the fractional and integer parts of the variable phase. The ADPLL prototype driving a three-stage transformer-coupled neutralized power amplifier delivers +5 dBm into a 50-Ω load, while consuming 89 mW (ADPLL: 48 mW and PA: 41 mW) from a single 1.2-V supply. As the first 60-GHz ADPLL reported in literature, it achieves good phase noise (−15.8 dBc integrated PN), fast settling (3 μs), low reference spur levels (−74 dBc), making it an attractive replacement of analog charge-pump PLLs for mm-wave frequency synthesis.

To further improve the integrated PN in order to meet more stringent PN requirements for other mm-wave applications (e.g., IEEE802.11ad with 16QAM modulation at 60-GHz band [40]), a TDC with finer resolution (e.g., 4 ps) would be required to lower the in-band PN. Consequently, the in-band PN will be reduced by $20\log_{10}(12.2\text{ ps}/4\text{ ps}) = 9.7$ dB and the loop bandwidth can also be widened to further suppress the PN of the DCO and reduce the integrated PN at synthesizer output. The TDC resolution of 4 ps can be easily obtained in 65-nm CMOS by employing the high-resolution TDC techniques described in Section 6.5, for example, the vernier delay line [26,27], a GRO [28,29], a two-step TDC combing coarse and fine [30,31], or an interpolation-based TDC [32].

REFERENCES

[1] W. Wu, X. Bai, R.B. Staszewski, J.R. Long, A 56.4-63.4 GHz spurious free all-digital fractional-N PLL in 65 nm CMOS, IEEE Int. Solid-State Circuits Conf. Dig. Tech Papers, Feb. 2013, pp. 352−353.

[2] A.G. Stove, Linear FMCW radar techniques, IEE Proc. F: Radar Signal Process. 139 (5) (1992) 343−350.

[3] G.M. Brooker, Understanding millimeter wave FMCW radars, Proc. Int. Conf. on Sensing Tech., Nov. 2005, pp. 152–157.
[4] T. Mitomo, N. Ono, H. Hoshino, Y. Yoshihara, O. Watanabe, I. Seto, A 77 GHz 90 nm CMOS transceiver for FMCW radar applications, IEEE J. Solid-State Circuits 45 (4) (2010) 928–937.
[5] Y.-A. Li, M.-H. Hung, S.-J. Huang, J. Lee, A fully integrated 77 GHz FMCW radar system in 65 nm CMOS, IEEE Int. Solid-State Circuits Conf. Dig. Tech. Papers, Feb. 2010, pp. 216–217.
[6] H. Sakurai, Y. Kobayashi, T. Mitomo, O. Watanabe, S. Otaka, A 1.5 GHz-modulation-range 10 ms-modulation-period 180 kHz rms-frequency-error 26 MHz-reference mixed-mode FMCW synthesizer for mm-wave radar application, IEEE Int. Solid-State Circuits Conf. Dig. Tech. Papers, Feb. 2011, pp. 292–293.
[7] C. Wagner, A. Stelzer, H. Jager, PLL architecture for 77-GHz FMCW radar systems with highly-linear ultra-wideband frequency sweeps, IEEE Int. Microwave Symp. Dig., June 2006, pp. 399–402.
[8] I. Komarov, S. Smolskiy, Fundamentals of Short-Range FM Radar, Artech House, 2003.
[9] W. Wu, J.R. Long, R.B. Staszewski, A digital ultra-fast acquisition linear frequency modulated PLL for mm-wave FMCW radars, Proc. IEEE Radio Frequency Integrated Technology Symp., Dec. 2009, pp. 32–35.
[10] K. Pourvoyeur, R. Feger, S. Schuster, A. Stelzer, L. Maurer, Ramp sequence analysis to resolve multi target scenarios for a 77-GHz FMCW radar sensor, Proc. Int. Conf. on Information Fusion, June 2008, pp. 1–7.
[11] K. Scheir, G. Vandersteen, Y. Rolain, P. Wambacq, A 57-to-66 GHz quadrature PLL in 45 nm digital CMOS, IEEE Int. Solid-State Circuits Conf. Dig. Tech. Papers, Feb. 2009, pp. 494–495.
[12] C. Lee, S. Liu, A 58-to-60.4 GHz frequency synthesizer in 90 nm CMOS, IEEE Int. Solid-State Circuits Conf. Dig. Tech. Papers, Feb. 2007, pp. 196–197.
[13] A. Musa, R. Murakami, T. Sato, W. Chaivipas, K. Okada, A. Matsuzawa, A low phase noise quadrature injection locked frequency synthesizer for mm-wave applications, IEEE J. Solid-State Circuits 46 (11) (2011) 2635–2649.
[14] W.F. Egan, Phase Lock Basics, Wiley, New York, NY, 1998.
[15] R.B. Staszewski, K. Waheed, F. Dulger, O.E. Eliezer, Spur-free multirate all-digital PLL for mobile phones in 65 nm CMOS, IEEE J. Solid-State Circuits 46 (12) (2011) 2904–2919.
[16] R.B. Staszewski, J.L. Wallberg, S. Rezeq, C.-M. Hung, O.E. Eliezer, S.K. Vemulapalli, et al., All-digital PLL and transmitter for mobile phones, IEEE J. Solid-State Circuits 40 (12) (2005) 2469–2482.
[17] L. Vercesi, L. Fanori, F. De Bernardinis, A. Liscidini, R. Castello, A dither-less all digital PLL for cellular transmitters, IEEE J. Solid-State Circuits 47 (8) (2012) 1908–1920.
[18] G. Marzin, S. Levantino, C. Samori, A.L. Lacaita, A 20 Mb/s phase modulator based on a 3.6 GHz digital PLL with −36 dB EVM at 5 mW power, IEEE J. Solid-State Circuits 47 (12) (2012) 2974–2988.
[19] R.B. Staszewski, J. Wallberg, Apparatus and Method for Acquisition and Tracking Bank Cooperation in a Digitally Controlled Oscillator. US Patent No. 7,746,185. June 2010.
[20] EMX User's Manual, Integrand Software, Inc., 2011, Berkeley Heights, NJ.
[21] P. Heydari, R. Mohanavelu, Design of ultrahigh-speed low-voltage CMOS CML buffer and latches, IEEE Trans. Very Large Scale Integration (VLSI) Syst. 12 (10) (2004) 1081–1093.

[22] D. Lim, J. Kim, J.-O. Plouchart, C. Cho, D. Kim, R. Trzcinski, et al., Performance variability of a 90 GHz static CML frequency divider in 65 nm SOI CMOS, IEEE Int. Solid-State Circuits Conf. Dig. Tech. Papers, Feb. 2007, pp. 542–543.
[23] L. Li, P. Reynaert, M. Steyaert, A 60 GHz 15.7 mW static frequency divider in 90 nm CMOS, Proc. European Solid-State Circuits Conf., Sept. 2010, pp. 246–249.
[24] R.B. Staszewski, S. Vemulapalli, P. Vallur, J. Wallberg, P.T. Balsara, 1.3 V 20 ps time-to-digital converter for frequency synthesis in 90-nm CMOS, IEEE Trans. Circuits Syst. II Express Briefs 53 (3) (2006) 220–224.
[25] B.N. Nikolic, V.G. Oklobdzija, V. Stajonovic, W. Jia, J. Chiu, M. Leung, Improved sense-amplifier-based flip-flop: design and measurements, IEEE J. Solid-State Circuits 35 (6) (2000) 876–884.
[26] P. Dudek, S. Szczepanski, J.V. Hatfield, A high-resolution CMOS time-to-digital converter utilizing a vernier delay line, IEEE J. Solid-State Circuits 35 (2) (2000) 240–247.
[27] L. Vercesi, L. Fanori, F. De Bernardinis, A. Liscidini, R. Castello, A dither-less all digital PLL for cellular transmitters, IEEE J. Solid-State Circuits 47 (8) (2012) 1908–1920.
[28] C.-M. Hsu, M.Z. Straayer, M.H. Perrott, A low-noise, wide-BW 3.6 GHz digital $\Delta\Sigma$ fractional-N frequency synthesizer with a noise-shaping time-to-digital converter and quantization noise cancellation, IEEE Int. Solid-State Circuits Conf. Dig. Tech. Papers, Feb. 2008, pp. 340–617.
[29] P. Lu, A. Liscidini, P. Andreani, A 3.6 mW, 90 nm CMOS gated-vernier time-to-digital converter with an equivalent resolution of 3.2 ps, IEEE J. Solid-State Circuits 47 (7) (2012) 1626–1635.
[30] M. Lee, A.A. Abidi, A 9 b, 1.25 ps resolution coarse–fine time-to-digital converter in 90 nm CMOS that amplifies a time residue, IEEE J. Solid-State Circuits 43 (4) (2008) 769–777.
[31] C.-S. Hwang, P. Chen, H.-W. Tsao, A high-precision time-to-digital converter using a two-level conversion scheme, IEEE Trans. Nucl. Sci. 51 (4) (2004) 1349–1352.
[32] S. Henzler, S. Koeppe, W. Kamp, H. Mulatz, D. Schmitt-Landsiedel, 90 nm 4.7 ps-resolution 0.7-LSB single-shot precision and 19 pJ-per-shot local passive interpolation time-to-digital converter with on-chip characterization, IEEE Int. Solid-State Circuits Conf. Dig. Tech. Papers, Feb. 2008, pp. 548–549.
[33] M. Lee, M.E. Heidari, A.A. Abidi, A low-noise wideband digital phase-locked loop based on a coarse–fine time-to-digital converter with subpicosecond resolution, IEEE J. Solid-State Circuits 44 (10) (2009) 2808–2816.
[34] I. Bashir, R.B. Staszewski, O. Eliezer, B. Banerjee, P.T. Balsara, A novel approach for mitigation of RF oscillator pulling in a polar transmitter, IEEE J. Solid-State Circuits (JSSC) 46 (2) (2011) 403–415.
[35] W. Wu, R.B. Staszewski, J.R. Long, A 56.4-to-63.4 GHz multi-rate all-digital fractional-N PLL for FMCW radar applications in 65-nm CMOS, IEEE J. Solid-State Circuits 49 (5) (2014) 1081–1096.
[36] K. Scheir, G. Vandersteen, Y. Rolain, P. Wambacq, A 57-to-66 GHz quadrature PLL in 45 nm digital CMOS, IEEE Int. Solid-State Circuits Conf. Dig. Tech. Papers, Feb. 2009, pp. 494–495.
[37] A. Musa, R. Murakami, T. Sato, W. Chaivipas, K. Okada, A. Matsuzawa, A low phase noise quadrature injection locked frequency synthesizer for mm-wave applications, IEEE J. Solid-State Circuits 46 (11) (2011) 2635–2649.

[38] X. Yi, C.C. Boon, H. Liu, J.F. Lin, J.C. Ong, W.M. Lim, A 57.9-to-68.3 GHz 24.6 mW frequency synthesizer with in-phase injection-coupled QVCO in 65 nm CMOS, IEEE Int. Solid-State Circuits Conf. Dig. Tech. Papers, Feb. 2013, pp. 354–355.
[39] C.-C. Hung, S.-I. Liu, A 40-GHz fast-locked all-digital phase-locked loop using a modified bang-bang algorithm, IEEE Trans. Circuits Syst. II Express Briefs 58 (6) (2011) 321–325.
[40] IEEE 802.11ad. (2012). Part 11: Wireless LAN Medium Access Control (MAC) and Physical Layer (PHY) Specifications Amendment 3: Enhancements for Very High Throughput in the 60 GHz Band.

CHAPTER 7

Digital Techniques for Higher RF Performance

Contents

Previous chapters have introduced the all-digital PLL architecture and described the design of key circuit building blocks including: millimeter (mm)-wave DCO, high-frequency divider chain, TDC, and reference slicer. When striving for higher RF performance, its stability over PVT and ultimately high manufacturing yield, calibration techniques are essential; hence they are the focus of this chapter.

The digitally intensive architecture, by its own nature, facilitates the various digital calibration techniques in order to improve the synthesizer performance stability. The major calibrations required in a high-performance ADPLL consist of a DCO gain calibration for the tracking bank (previously discussed in Section 4.2), TDC gain calibration (elaborated on in Section 6.5), and calibrations related to the wideband frequency modulation. Successful execution of the DCO and TDC gain calibrations allows to establish the desired closed-loop transfer function of the PLL, thus precisely controlling its loop bandwidth. This is crucial for many applications. For a wideband linear frequency modulation required by FMCW radar applications, not only the DCO gain of the fine tuning banks needs to be calibrated, but also its tuning characteristics (i.e., f_o vs. tuning code) over multiple tuning banks must be determined and

DOI: http://dx.doi.org/10.1016/B978-0-12-802207-8.00007-1

linearized in order to achieve linear frequency modulation across several GHz. The frequency step-size mismatch within the fine-tuning bank is also critical when striving for ultra-linear FM sweeps. These techniques will be elaborated on in this chapter using the 60-GHz ADPLL as a design example.

7.1 FREQUENCY TUNING NONLINEARITY IN A MULTIBANK DCO

The 60-GHz DCO prototype elaborated on in Chapter 5 achieves a multi-GHz tuning range and ~2-MHz raw resolution or step size. Ideally, a single tuning bank with a constant K_{DCO} across the modulation range would be employed to do the job. However, for practical reasons, the DCO tuning must be segmented into coarse-tuning (CB), mid-coarse tuning (MB), and fine-tuning banks (FB) (i.e., each with different K_{DCO}) to realize both high resolution and a wide tuning range. Consequently, the wideband triangular modulation traverses through all these three tuning banks, as shown in Figure 7.1b. Integer bits in FB_{Mod} are matched to each other within 5% by employing the "virtual" dummy bank configuration earlier illustrated in Figure 6.5. Tuning step mismatches of 15% in MB and CB (shown in Figures 8.11 and 8.12) are much larger than in FB because dummy cells are not employed for coarse-tuning. Moreover, the K_{DCO} in MB and FB varies with the CB tuning word due to the

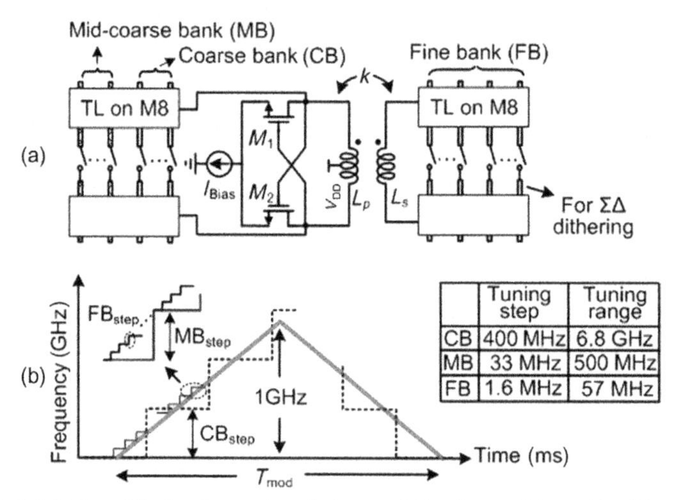

Figure 7.1 (a) Simplified schematic of the multibank 60-GHz DCO; (b) wideband triangular modulation traversing all three tuning banks.

ultra-wide 7-GHz coarse-tuning range (see Figures 8.12 and 8.13). When the capacitance increases by ΔC, the oscillation frequency f_o will decrease by Δf, or approximately $\frac{1}{2}\frac{\Delta C}{C_o}f_o$ (derived from Eq. 3.2). Therefore, Δf will vary with f_o even for the same ΔC, which is the case for our modulation frequency range of a few gigahertz.

In addition, mismatches between the fractional bits (for $\Sigma\Delta$ dithering) and the integer bits in the fine-tuning bank are large because the three fractional bits in FB are located at the edge of the tuning bank (Figure 7.1a). These mismatches result in a frequency error in the ramp generation and can degrade ramp linearity. Therefore, nonlinearities in the DCO tuning curve must be calibrated and compensated in real-time in order to implement the linear triangular FM with wide modulation range (Figure 7.1b).

7.2 MULTIBANK DCO GAIN CALIBRATION AND LINEARIZATION

The dependence of DCO oscillator gain, K_{DCO}, on PVT and oscillation frequency requires estimation of the gain on an as-needed basis during the actual operation. Traversing multiple banks, each with its distinct tuning characteristics, is unavoidable for the multi-GHz modulation required by the FMCW radar applications and many other mm-wave applications. As discussed in Section 6.2, wideband two-point modulation relies on the accurate knowledge of K_{DCO}. Therefore, the DCO gain calibration and linearization techniques are essential for the generation of a linear frequency modulation.

Published DCO gain calibration and linearization techniques for low-GHz ADPLLs include: (i) digital normalization that measures the tuning word response to a training sequence of FM modulation to estimate the K_{DCO} of a linear fine-tuning bank [1], (ii) adaptive gain compensation by a sign-LMS loop that observes a phase error reaction to an FM stimulus [2,3], (iii) DCO FB mismatch characterization via an open-loop method [4], and (iv) signal predistortion in the direct modulation path by polynomial fitting to correct for the nonlinearity in FB [5]. The digital normalization algorithm has been described in Section 4.3. By executing the calculation algorithm just-in-time at the beginning of every wireless communication packet, the DCO gain can be conveniently tracked and compensated. The sign-LMS loop provides background regulation of the DCO gain or the gain ratio between various banks, but is difficult to apply here since each bit in CB and MB has a different K_{DCO}, and the K_{DCO} in FB also varies with CB settings as explained above. It requires over 10 K_{DCO} values to be calibrated in the background using an adaption algorithm that is rather complicated to implement and may not converge to a stable solution. Correcting K_{DCO} via a look-up table for individual bits in each bank employing open-loop calibration algorithms [4] requires a long calibration time (up to hours) and an unacceptably large look-up table for a gigahertz-range frequency modulation.

As an alternative, a fast, closed-loop DCO gain linearization technique for FMCW generation will be described in this section [6]. For a triangular modulation of a slope $k_{mod} = 2BW/T_{mod}$ (BW is the modulation extent and T_{mod} is the period of the triangular modulation, as shown in Figure 7.1b), the output frequency change within each modulation clock (CKM) is k_{mod}/f_{CKM}, where f_{CKM} is the modulation clock sampling rate. Instead of finding and storing accurate DCO tuning words (OTWs) for each frequency along the triangular modulation trajectory, accurate OTWs are determined only in the vicinity of the bank-switchover points (Figure 7.2). Between the two adjacent bank-switchover points, only FB is used for modulation, which is sufficiently linear, and one normalized K_{DCO} for each subrange is employed. Thus, the size of the look-up

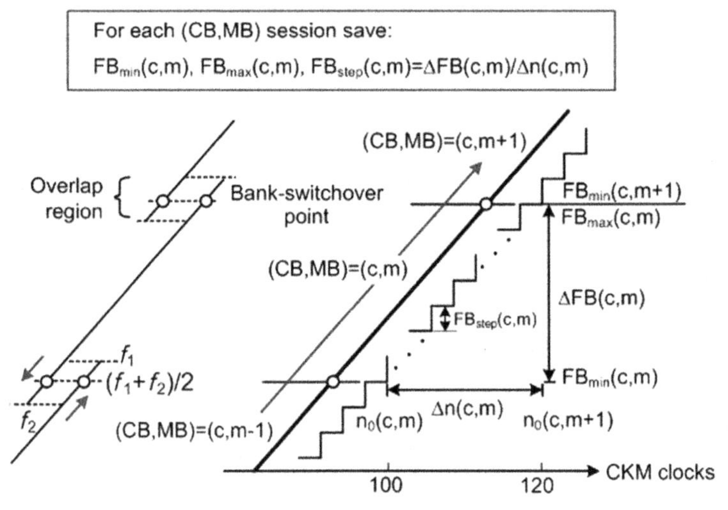

Figure 7.2 Principle of the multibank DCO gain calibration and linearization technique.

table is determined by the number of bank-switchover points and is independent of the FM rate and range. To ensure monotonic tuning against PVT, the total FB tuning range is set to 1.7× the frequency step size in CB. Thus, bank switchover may be performed at any frequency located in the overlap region shown in Figure 7.2. The mid-point of the overlap region is chosen for a robust switchover. When the upper and lower boundaries of the tuning word in FB are determined for CB = c and MB = m (i.e., $FB_{max}(c, m)$ and $FB_{min}(c, m)$), the normalized tuning step $FB_{step}(c, m)$ for each CKM is calculated by $FB_{step}(c, m) = \Delta FB(c, m)/\Delta n(c, m)$. Parameter $\Delta FB(c, m)$ is the frequency range for CB = c and MB = m, and $\Delta n(c, m)$ is the number of CKM cycles needed to modulate across $\Delta FB(c, m)$ for a specific chirp slope, k_{mod}. Thus, only three variables: $FB_{max}(c, m)$, $FB_{min}(c, m)$, and $FB_{step}(c, m)$ need to be saved in SRAM for each index (c, m). Note that two sets of DCO tuning words are saved for each switchover point to implement hitless modulation, for example, $FB_{max}(c, m)$ and $FB_{min}(c, m + 1)$. The required calibration time at power-up is reduced to only 4 s.

7.3 MISMATCH CALIBRATION OF THE FINE-TUNING BANK

The mismatch of the fractional tuning bit in FB_{Mod} (first-order $\Sigma\Delta$) with respect to the average K_{DCO} of the integer bits is characterized in an open-loop manner by forced on/off toggling of the fractional bit. Since small capacitance fluctuations in the DCO tank result in proportional frequency fluctuations, changes in capacitance, resulting from on/off switching of the dithering bit are evaluated by subtracting frequency measurements performed at each of the two states. This open-loop configuration is used since each toggling procedure addresses a *specific* fine-tuning bit, which could not be guaranteed through the normal modulation operation of the ADPLL. The frequency measurements are based on a counter within the ADPLL and multiple readings of the counter (e.g., M readings) that are averaged to reduce the quantization error in a single measurement of frequency deviation, especially in the presence of DCO phase noise, as shown in Figure 7.3. The tuning step value for this particular bit i is Δf_i, and can be calculated by

$$\Delta f_i = \frac{1}{M}\left[\sum_{i=1}^{M} f_{1_k} - \sum_{i=1}^{M} f_{0_k}\right]. \tag{7.1}$$

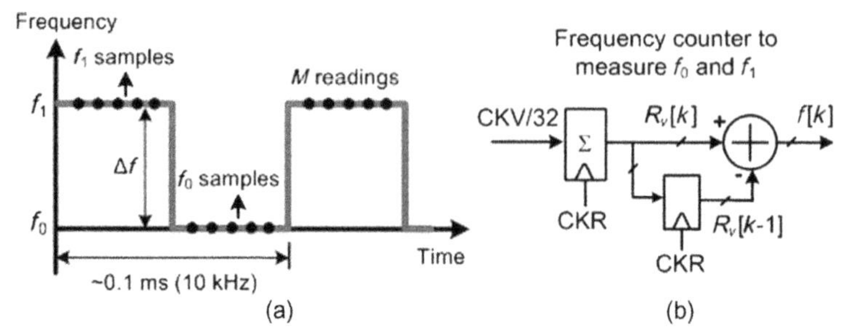

Figure 7.3 Open-loop DCO gain calibration based on toggling.

The number of measurements (N) used is typically on the order of 2^{15} for a 1% fine-tuning step accuracy. The resulting frequency tuning step after averaging is

$$\Delta f_{avg,i} = \frac{1}{N}\sum_{i=1}^{N}\Delta f_i. \tag{7.2}$$

Thus, the normalized tuning step mismatch between the fractional bit and the integer bit i of the FB_{Mod} is

$$\varepsilon_i = \frac{\Delta f_{avg,i} - \Delta f_{FB}}{\Delta f_{FB}}, \tag{7.3}$$

where, Δf_{FB} is the estimated frequency tuning step of the integer bits of the FB_{Mod}, using the method described in Section 4.2.

7.4 SYNCHRONIZATION IN A MULTIRATE SYSTEM

Figure 7.4 elaborates on the multirate two-point FM in the 60-GHz ADPLL-based FMCW transmitter. As explained in Chapter 6, the direct modulation path operates at a high clock rate (f_{CKM}), which is a down-divided DCO clock (by a small integer) to obtain high sweep linearity. The modulated DCO output frequency step in each modulation clock (CKM) is $k_{\text{mod}}/f_{\text{CKM}}$. For this reason, CKM is configurable from CKV/128 (~450 MHz) to CKV/1024 (~56 MHz) to minimize power consumption according to the required modulation ramp slope (k_{mod}). The compensation path is applied to the frequency reference and so it operates at the retimed reference clock (CKR) rate, f_{R}. During FM, CKV varies piecewise-linearly with time and so does CKM.

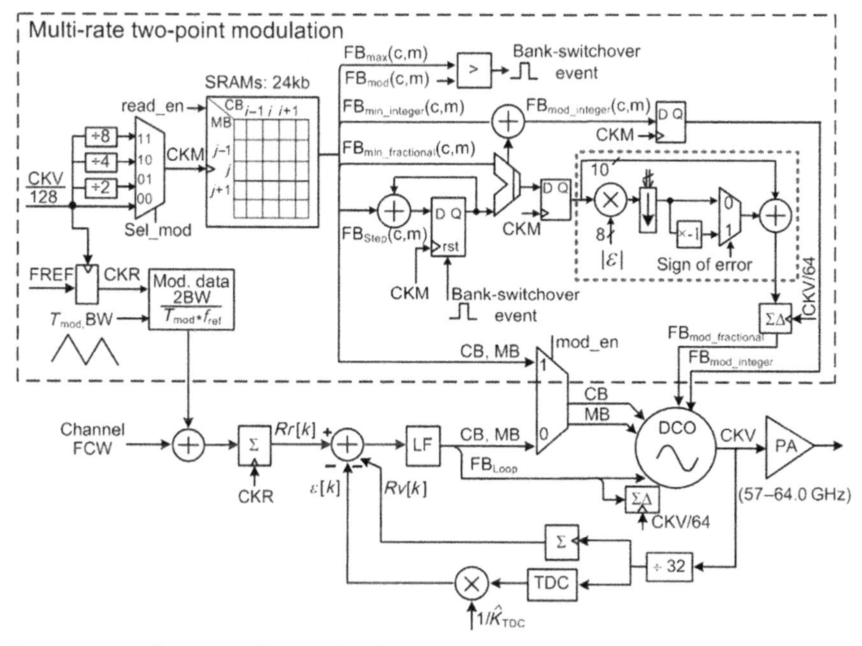

Figure 7.4 Multirate modulation for the 60-GHz ADPLL-based FMCW transmitter.

The two functional parts of the ADPLL-based frequency modulator, that is, the phase error calculator and the data modulator, have their own separate clock domains: FREF and CKV, respectively. Since their frequency relationship is a time-varying fractional number, their interfaces normally would require sampling rate converters. However, this is not necessary in this architecture because the system clock CKR is always synchronized with the modulator clock CKM via re-sampling of FREF by CKV/128. Furthermore, the fine-tuning bank used for data modulation (FB_{Mod}) is physically separated from the FB_{Loop}, which is used for phase error corrections. Regarding the coarse-tuning banks CB and MB, they are controlled by the direct datapath upon modulation (Figure 7.4). Therefore, no sampling rate conversion is required for the DCO tuning word.

During the modulation, a state-machine controls the access to SRAMs and reads out the proper data before the bank switchover. Operation at high speed is simplified to an accumulation of the FB_{step} and a comparator to generate the bank switchover event. Meanwhile, a frequency step equal to $k_{mod}/f_R/32$ (32 is the prescaler's division ratio in the feedback loop) is added as compensation to the frequency reference at every CKR to obtain the wideband FM output.

The mismatch of the dithering bit in FB is obtained via the open-loop calibration described in Section 7.3, and compensated in the direct path using the logic highlighted by the dotted line in Figure 7.4. The compensation mechanism is based on a digital gain correction factor that is applied to the 10-bit FB_{Mod} fractional tuning word before it is fed to the fractional tuning unit, where it is converted into the appropriate dithering signal. The correction factor can be represented as a sum of one and the normalized mismatch error (ε). Accordingly, it is implemented using a reduced-size multiplier followed by an adder. The magnitude of the error ε has been limited to 8 bits, allowing for a dynamic range of mismatch errors up to 25% and a theoretical resolution of 0.1%, which is more than sufficient. In addition, the fractional unit in FB_{Mod} (Figure 7.1a) is sized in this design so that the compensating factor is always a fractional number, thereby avoiding potential overflow.

7.5 EXPERIMENTAL RESULTS OF FMCW TRANSMITTER

The above digital calibration techniques are implemented on the same 60-GHz ADPLL prototype, which was shown in Figure 6.25 to obtain the gigahertz-level linear frequency modulation for FMCW radar applications [6,7]. The frequency modulation measurement results are shown in this section to demonstrate the high RF performance achieved by employing the above calibration techniques.

7.5.1 FSK Modulation

A two-point modulation at the FREF rate, employing only the DCO fine-tuning bank is demonstrated by FSK modulating the 60-GHz carrier at a rate of 50 kHz with a maximum frequency deviation of 40 MHz. The DCO gain for FB_{Mod} and the TDC gain are calibrated automatically via the digital averaging techniques described in Sections 4.3 and 6.5. The calibrated K_{DCO} is then applied to the gain normalization multiplier in the direct modulation path (see Figure 6.1). The demodulated signal is measured using a Rohde & Schwarz FSUP signal source analyzer with FM demodulation firmware, and the waveform measured at the CKV/32 output is shown in Figure 7.5a. The sharp transition edges in the step-response (which require many harmonics) confirm the wideband FM capability and demonstrate the effectiveness of the built-in K_{DCO} calibration. The $f_R/\hat{K}_{DCO}$ multiplier is then

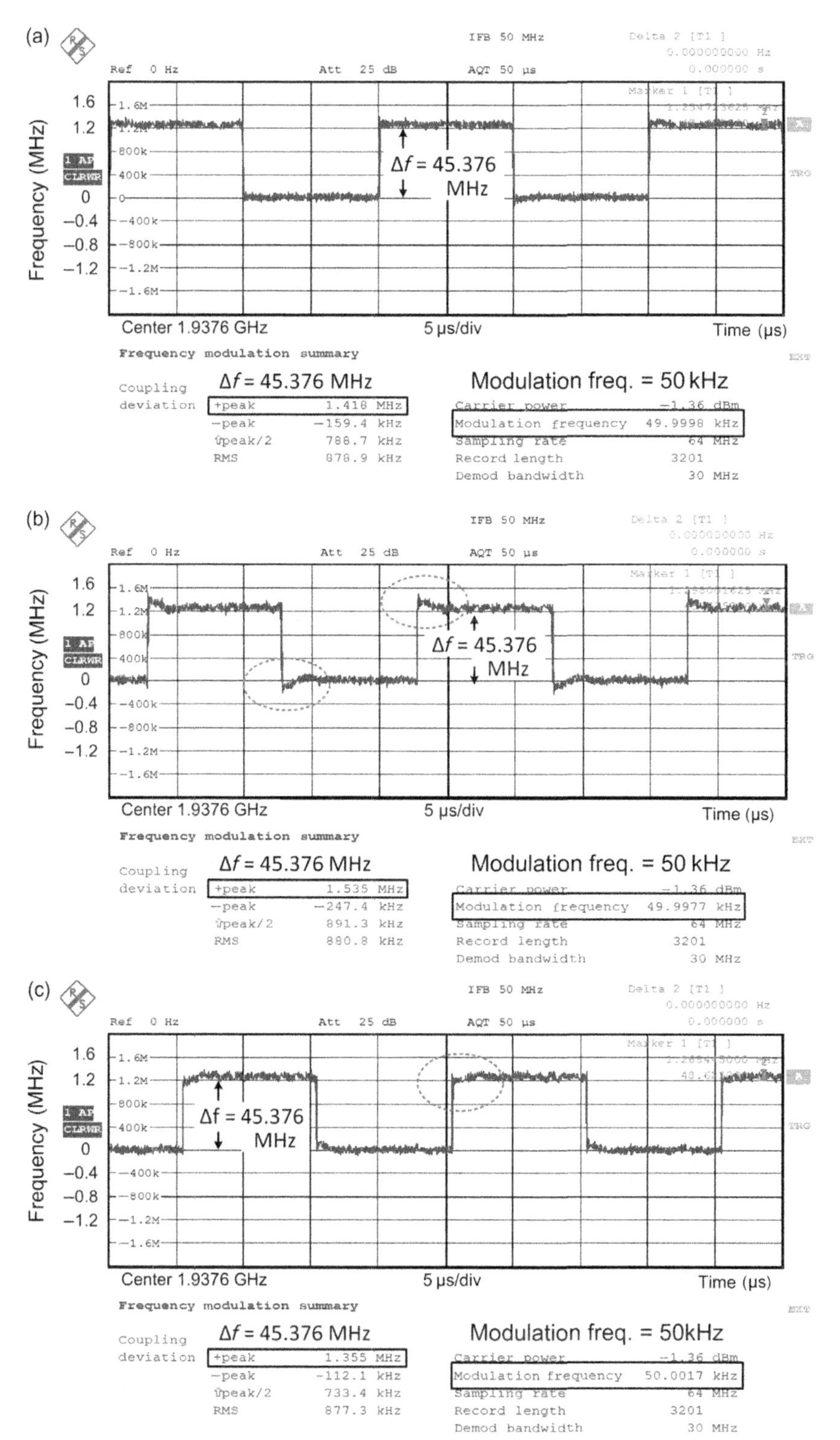

Figure 7.5 Measured demodulated signal at CKV/32 during FSK: (a) using autocalibrated K_{DCO}, (b) multiplier factor is larger than optimal, and (c) multiplier factor is smaller than optimal.

perturbed intentionally from the self-calibrated optimum to demonstrate the effect of an incorrect K_{DCO} estimation. It can be seen from Figure 7.5b that a larger multiplier factor introduces strong overshoot in the demodulated signal rising/falling edges, while a waveform with an "undershoot" is observed for a smaller multiplier in Figure 7.5c.

7.5.2 FMCW Modulation

To generate an FMCW chirp with high sweep linearity, a multirate two-point modulation is adopted, in which the modulation paths operate at a higher rate than FREF (i.e., programmable from CKV/128 to CKV/1024). All three tuning banks (CB, MB, and FB) are traversed when the modulation range is wider than 300 MHz. The closed-loop, multibank DCO gain linearization technique presented in Section 7.2 is performed automatically prior to modulation, and the resulting look-up table is stored in 24-kbit SRAMs, all accomplished within 4 s. Both slow and fast modulation slopes are used to characterize the chirp linearity by programming various values for the modulation range (BW) and period (T_{mod}). Figure 7.6a shows the FMCW chirp spectrum measured at the PA output for a 1.22-GHz modulation range centered at 62.1 GHz. The T_{mod} is 8.2 ms, forming a slow, triangular chirp. CKM is configured at CKV/1024 to reduce the power consumption of the digital part by 20% compared to that while operating at CKV/128. The instantaneous output frequency of the FMCW synthesizer and the frequency error compared to an ideal chirp are also plotted in Figure 7.6b. The root-mean-square (rms) frequency error is only 117 kHz. Figure 7.7a shows the result when the modulation slope is 4× faster than the case in Figure 7.6b. The rms error in the frequency chirp is 148 kHz. For an ultra-fast chirp (1-GHz change in 210 μs, plotted in Figure 7.7b) the frequency error degrades to 384 kHz_{rms}. For the fast chirp, the direct modulation path operates at a higher CKM (e.g., CKV/128 = ~500 MHz), and thus the accumulator used in the direct path to calculate the tuning word of FB_{Mod} has less bits compared to a slow chirp in order to operate at a higher speed. This reduces the resolution in the direct path. Meanwhile, the maximum sampling rate available during the measurement data collection is 64 MHz, which increases measurement inaccuracy for a faster chirp (e.g., 1 GHz/210 μs/64 MHz = 74.4 kHz for a triangular chirp of $BW = 1$ GHz and $T_{mod} = 420$ μs).

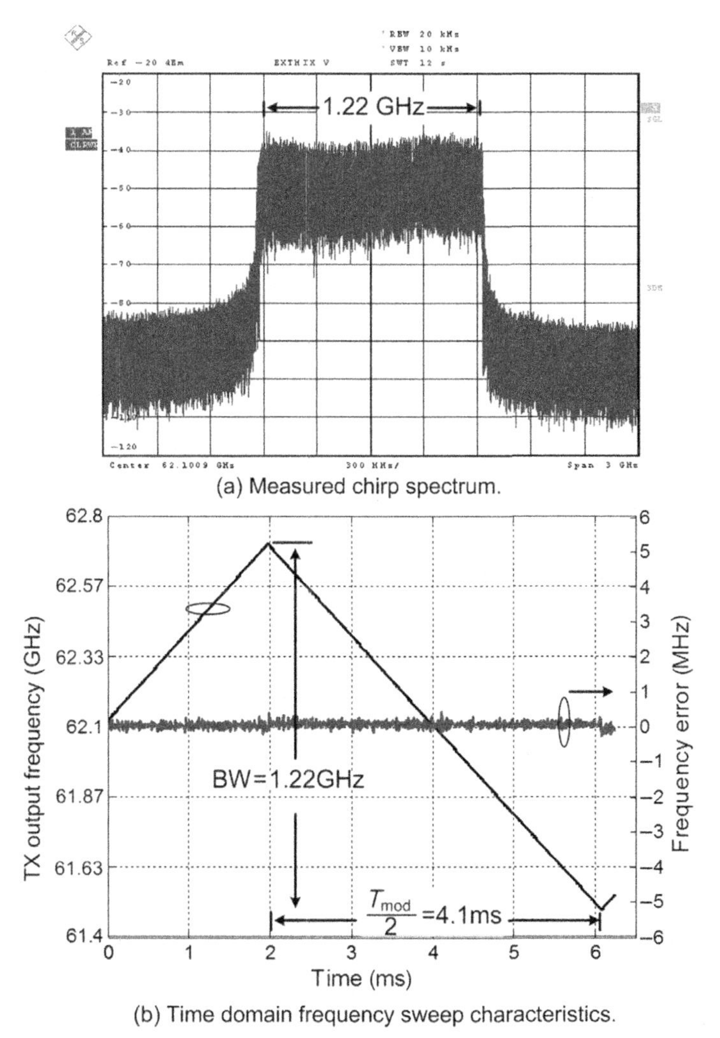

(a) Measured chirp spectrum.

(b) Time domain frequency sweep characteristics.

Figure 7.6 (a) Measured chirp spectrum and (b) time-domain frequency characteristics of the chirp (for $BW = 1.22$ GHz and $T_{mod} = 8.2$ ms).

The performance of the 60-GHz all-digital FMCW synthesizer is summarized in Table 7.1. Compared to state-of-the-art FMCW generators, the all-digital architecture achieves wider modulation range for varying modulation slopes, and better phase noise with lower power consumption.

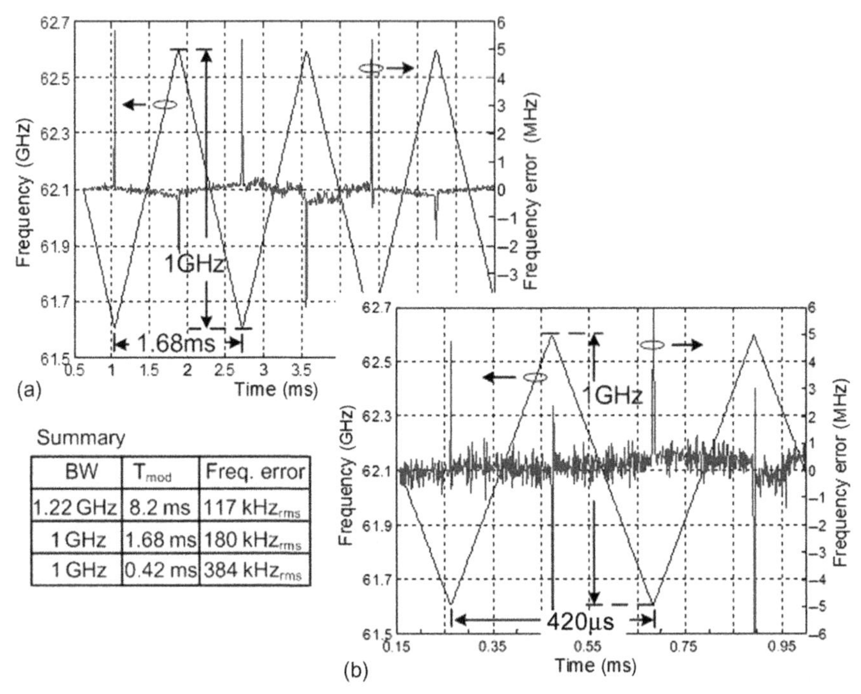

BW	T_{mod}	Freq. error
1.22 GHz	8.2 ms	117 kHz_{rms}
1 GHz	1.68 ms	180 kHz_{rms}
1 GHz	0.42 ms	384 kHz_{rms}

Figure 7.7 Measured frequency characteristics of FMCW signal for (a) $T_{mod} = 1.68$ ms, $BW = 1$ GHz; (b) $T_{mod} = 0.42$ ms, BW 1 GHz.

7.6 SUMMARY

Digital calibration techniques applied in a mm-wave ADPLL that help to achieve wideband, linear frequency modulation across several gigahertz range are discussed in this chapter. These techniques are implemented in the 60-GHz fractional-N ADPLL-based transmitter prototype in 65-nm CMOS that is capable of multirate, two-point FM. To execute ultra-linear FM, the ADPLL can calibrate and linearize the multibank DCO tuning curve to less than 10 kHz of frequency granularity within 4 s. It is further assisted by an open-loop DCO gain calibration algorithm based on toggling in order to obtain the mismatch between fractional bits and the integer bits of fine-tuning bank. This cannot be realized through the normal modulation capability of the ADPLL since the factional bits cannot be arbitrarily controlled during the closed-loop operation. The modulation path operates at the down-divided DCO frequency, which can be more than 10 times higher than the reference frequency (40 MHz) in order to reduce the instantaneous frequency error in chirp generation and

Table 7.1 FMCW synthesizer performance comparison

	This work	ISSCC (2011) [8]	JSSC (2010) [9]	ISSCC (2010) [10]
Architecture	ADPLL + multirate 2-point mod.	Mixed-mode	DDFS + PLL	Fractional-N
Frequency (GHz)	56.4−63.4	82.1−83.8	78.1−78.8	75.6−76.3
CMOS	65 nm	65 nm	90 nm	65 nm
Reference	40 MHz	26 MHz	77 MHz	700 MHz
Modulation slope (km)	Fast: 1 GHz/0.21 ms Slow: 1.22 GHz/4.1 ms	Fast: 1.5 GHz/1 ms Slow: 0.5 GHz/5 ms	700 MHz/0.22 ms	500 MHz/0.5 ms
Frequency error (r.m.s)	Fast: 384 kHz Slow: 117 kHz[a]	Fast: 179 kHz Slow: 170 kHz	1.05 MHz[a]	<300 kHz[a]
PN @ 1 MHz	−90 dBc/Hz	−84 dBc/Hz	−85 dBc/Hz	−84.6 dBc/Hz
Supply	1.2 V	1.2 V	1.2 V	1.2 V
P_{DC} (mW)	48 + 41 (PA)	152 with buffer	101 + 35 (TX)	73 + 115 (PA)
P_{our} (50 Ω)	+5 dBm	NA	−2.8 dBm	+10.5 dBm

[a]Includes turnaround point.

enable wideband modulation. The two clock domains are synchronized to avoid metastability. The high-speed modulation clock is programmable to optimize power consumption for various modulation schemes. The measured frequency error (ramp nonlinearity) of the FMCW signal is only 117 kHz (rms) for a 62-GHz carrier with 1.22-GHz bandwidth. These digital calibration techniques developed for FMCW generation can also be used in mm-wave ADPLLs to form other types of wideband linear frequency modulation and obtain high RF performance.

REFERENCES

[1] R.B. Staszewski, D. Leipold, P.T. Balsara, Just-in-time gain estimation of an RF digitally-controlled oscillator for digital direct frequency modulation, IEEE Trans. Circuits Syst. II 50 (11) (2003) 887–892.
[2] R.B. Staszewski, J. Wallberg, G. Feygin, M. Entezari, D. Leipold, LMS-based calibration of an RF digitally controlled oscillator for mobile phones, IEEE Trans. Circuits Syst. II Express Briefs 53 (3) (2006) 225–229.
[3] G. Marzin, S. Levantino, C. Samori, A.L. Lacaita, A 20 Mb/s phase modulator based on a 3.6 GHz digital PLL with −36 dB EVM at 5 mW power, IEEE J. Solid-State Circuits 47 (12) (2012) 2974–2988.
[4] O. Eliezer, R.B. Staszewski, J. Mehta, F. Jabbar, I. Bashir, Accurate self-characterization of mismatches in a capacitor array of a digitally-controlled oscillator, in: Proceedings of IEEE Dallas Circuits and Systems Workshop, October 2010, pp. 1–4.
[5] L. Vercesi, L. Fanori, F. De Bernardinis, A. Liscidini, R. Castello, A dither-less all digital PLL for cellular transmitters, IEEE J. Solid-State Circuits 47 (8) (2012) 1908–1920.
[6] W. Wu, X. Bai, R.B. Staszewski, J.R. Long, mm-Wave FMCW radar transmitter based on multi-rate ADPLL, in: IEEE Radio-frequency Integration Circuits Symposium (RFIC), June 2013, pp. 107–110.
[7] W. Wu, R.B. Staszewski, J.R. Long, A 56.4-to-63.4 GHz multi-rate all-digital fractional-N PLL for FMCW radar applications in 65-nm CMOS, IEEE J. Solid-State Circuits 49 (5) (2014) 1081–1096.
[8] H. Sakurai, Y. Kobayashi, T. Mitomo, O. Watanabe, S. Otaka, A 1.5 GHz-modulation-range 10 ms-modulation-period 180 kHzrms-frequency-error 26 MHz-reference mixed-mode FMCW synthesizer for mm-wave radar application, in IEEE International Solid-State Circuits Conference Digest of Technical Papers, vol. 40 (12), pp. 292–294, 2011.
[9] T. Mitomo, N. Ono, H. Hoshino, Y. Yoshihara, O. Watanabe, I. Seto, A 77 GHz 90 nm CMOS transceiver for FMCW radar applications, IEEE J. Solid-State Circuits 45 (4) (2010) 928–937.
[10] Y.-A. Li, M.-H. Hung, S.-J. Huang, J. Lee, A fully integrated 77 GHz FMCW radar system in 65 nm CMOS, in: IEEE International Solid-State Circuits Conference Digest of Technical Papers, vol. 43, pp. 216–217, 2010.

CHAPTER 8

Design for Test of the mm-Wave ADPLL

Contents

8.1 TESTING CHALLENGES FOR THE RF SYNTHESIZER

Testing and debugging of a complex, mixed-signal SoC for wireless consumer applications affect both product development time and cost. Testing costs are usually high, since high-volume wireless applications must be extensively checked for defects, RF performance, and wireless standards compliance with sophisticated and expensive test equipment. Debugging becomes more difficult because of the constrained number of I/Os available to access an SoC during testing.

Frequency synthesizers and transmitters are tested for RF performance by measuring the carrier frequency, phase noise spectral density, integrated phase noise, spurious tone content, modulated spectrum, and modulated phase error trajectory at the RF output when modulation stimuli are applied [1,2]. These RF performance tests are conducted during a development phase as well as in production. When debugging, it is rather difficult to identify what causes a failure of the PLL to lock and how to fix the problem, or to determine why performance is degraded at

Millimeter-Wave Digitally Intensive Frequency Generation in CMOS.
DOI: http://dx.doi.org/10.1016/B978-0-12-802207-8.00008-3

the RF output (e.g., poor phase noise or high spur levels). Although functional testing of individual blocks can be conducted open-loop, relating closed-loop PLL performance to the performance of individual circuit blocks is non-trivial due to the tight feedback nature of a PLL [3].

Alternatively, built-in self-test (BIST) requires no external test equipment. It is widely used for digital ICs to reduce their testing time and cost, while increasing their test coverage. Including BIST capabilities in mixed-signal RFICs would lessen the need for high-performance test equipment and provide data for debugging purposes [4]. These benefits are easier to realize on RFICs with an ADPLL, as most of the ADPLL circuitry and its signal path can be accessed and evaluated by digital signal processing with little hardware overhead. By comparison, charge-pump PLLs are difficult to adapt for BIST because loading sensitive analog nodes for test purposes changes their loop behavior and skews the measured data [1].

Aside from debugging and defect fault testing, the BIST has been applied to RF performance characterization of low-GHz ADPLLs, as reported in Refs. [5–8], where digital signal processing of a lower-frequency internal signal is used to ascertain RF performance without any external test equipment. An ADPLL is usually integrated in a digitally intensive SoC (although the 60-GHz ADPLL prototype demonstrated in Chapter 6 is a standalone chip), consisting of a digital baseband processor, SRAM memories, and power management functions. Due to the potential for reuse of SRAM and signal processing circuitry, very little hardware overhead is required to implement design-for-test (DFT) and design-for-characterization (DFC) techniques into an ADPLL. For example, the system snapshot can be triggered by a sequence of major internal events in normal operation of the PLL to capture the transient behavior of the loop at a particular moment for observation, and to provide a means of analyzing loop operation analogous to simulation methodologies.

In this chapter, we introduce a systematic approach to debugging and testing by employing system snapshots, as well as BIST and built-in self-characterization (BISC) for the ADPLL and an ADPLL-based transmitter operating at millimeter (mm)-wave frequencies. The DFT techniques are described in the context of a 60-GHz ADPLL-based FMCW radar transmitter (Figure 8.1), and the same concepts can be applied to digitally intensive PLLs and two-point modulation transmitters generally—especially those using a DCO, TDC, and digitization of a conventional phase/frequency detector output [9].

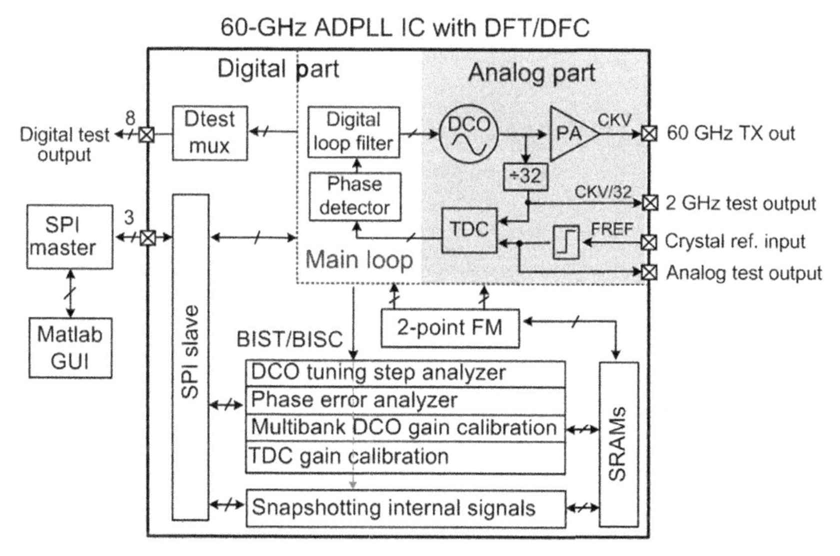

Figure 8.1 Block diagram of the 60-GHz ADPLL from a DFT/DFC perspective.

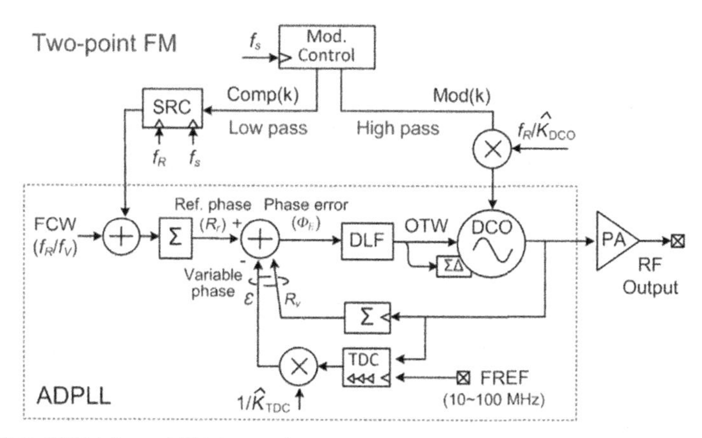

Figure 8.2 ADPLL-based FM transmitter.

8.2 CRITICAL SIGNALS IN ADPLL FOR DFT AND DFC

Figure 8.2 depicts a generic ADPLL-based frequency modulator architecture that has been described in detail in Section 6.2. The output frequency and phase are controlled by a negative feedback loop comprising the DCO, the TDC, and the incrementer to estimate the DCO phase $(R_v + \varepsilon)$, the frequency command word (FCW) accumulator to calculate

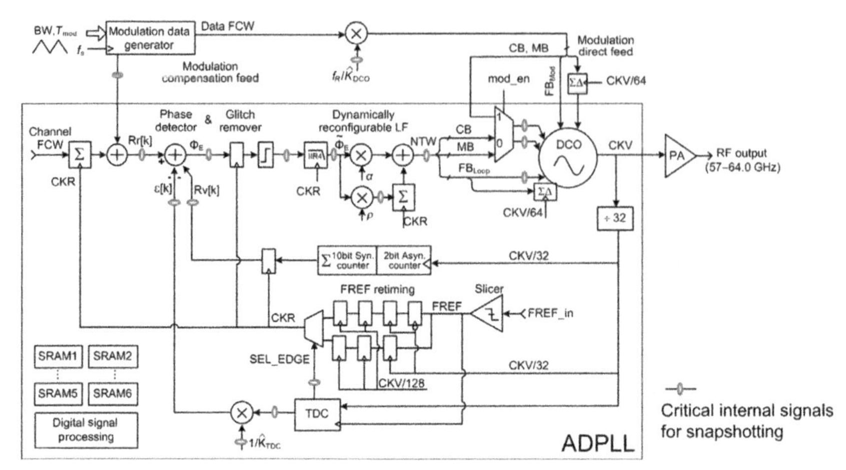

Figure 8.3 Block diagram of the 60-GHz ADPLL synthesizer.

the reference phase (R_r), the arithmetic subtractor to calculate the phase error (Φ_E), and the digital loop filter (LF) to control the ADPLL bandwidth and transfer function characteristics. The transmitter is configured to generate a linear frequency modulation by incorporating two-point FM directly into the ADPLL without the need for an up-conversion mixer. One data path modulates the DCO directly, while the other path compensates the frequency reference and prevents the modulating data from affecting Φ_E. Therefore, the variation of Φ_E indicates correlation with the RF performance measured at the transmitter output, which makes BIST possible.

A more detailed system diagram of the 60-GHz ADPLL prototype is shown in Figure 8.3. After the study of previous chapters, we should now be quite familiar with the circuit design and operation of this digitally intensive PLL. Now let us examine it from the test and debugging point of view. The 60-GHz ADPLL shown in Figure 8.3 includes several critical analog/mixed-signal building blocks (i.e., DCO, power amplifier, high-frequency divider, and TDC) and many key digital logic blocks (e.g., phase detector, LF, TDC/DCO gain calibration algorithm, two-point FM). Although the spectrum measured at the RF output (whether at CKV or CKV/32) indicates the PLL performance, it is difficult to locate the source of a problem from an unlocked spectrum or when many spurious tones are observed at the RF output. In those cases, internal digital signals clocked at the FREF rate (e.g., raw phase error Φ_E, filtered Φ_E, oscillator tuning word) should be monitored.

As shown in Figure 8.3, the phase error signal Φ_E is a numerical difference between the reference and variable phases at the digital output of the phase detector. Φ_E has a low-pass, unity-gain transfer to the variable phase at the ADPLL RF output (as described in Section 4.5). The response is flat up to the PLL bandwidth edge in type I or type II configurations with a sufficiently large damping factor (e.g., 1) [5]. Therefore, the trajectory of Φ_E correlates closely with the RF performance measured at the PLL output.

For a type II PLL, Φ_E, or its filtered version, has a zero mean once the loop is locked. Its variance represents the PN at the RF output with adequate accuracy. The trajectory of Φ_E reveals the transient behavior of the loop, loop stability, and its frequency response. For example, if there is an unwanted spurious tone in the RF spectrum, the tone frequency and its energy level can be sensed from spectral estimation of the Φ_E trajectory. For catastrophic errors during the ADPLL operation (e.g., the PLL loses lock), the filtered Φ_E is no longer flat and its variance exceeds normal operating bounds.

For two-point FM, the Φ_E trajectory may be different from the continuous-wave (CW) mode because the modulation data affect the Φ_E noise characteristics, as depicted in Figure 8.4 for a triangular modulation. Increased range of the Φ_E variation indicates greater phase noise in the system, which could be due to the DCO gain calibration inaccuracy or due to the digital-to-frequency conversion nonlinearity in the DCO modulation bank.

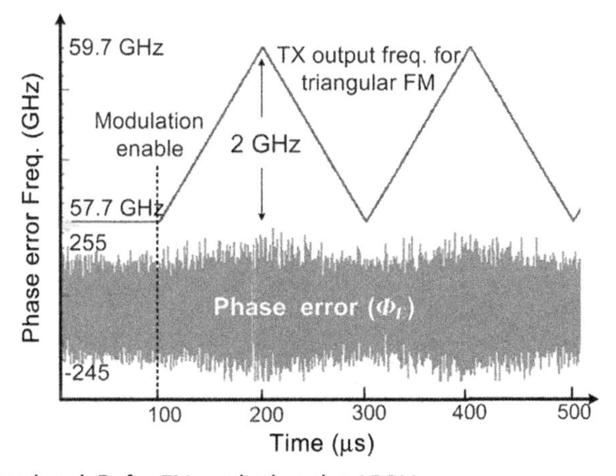

Figure 8.4 Simulated Φ_E for FM applied to the ADPLL.

Besides Φ_E, both the integer and fractional parts of the variable phase (R_v and ε), and reference phase (R_r) are informative for debugging, as $\Phi_E[k] = R_r[k] - (R_v[k] + \varepsilon[k])$. When tracing fractional spurs, observing $\varepsilon[k]$ is generally more effective. It reflects periodic behavior in the variable phase due to a TDC nonlinearity. In addition, other signals along the path from the phase detector to the DCO could also be used for DFT/DFC, such as internal signals in the LF, the scaled-down and filtered version of Φ_E at the LF output, and the DCO's oscillator tuning word (OTW). A frequency deviation of the DCO from its natural frequency can be ascertained by observing an output of the integral path accumulator when in type II operation. Alternatively, the output of the IIR filter, which is connected to the Φ_E, could be observed. The OTW can be in a binary form, or an encoded number that matches the DCO interface.

The aforementioned critical internal signals are intrinsic to any ADPLL, and are highlighted in Figure 8.3. Thanks to the digitally intensive nature of the ADPLL, these signals can be monitored as well as processed on-chip for fault testing and performance characterization. The following sections describe how the monitoring and on-chip signal processing are performed.

8.3 DFT TECHNIQUES FOR ADPLL

8.3.1 Snapshotting Internal Signals for DFT and DFC

Parallel outputs are normally used to monitor internal digital signals in real time. An on-chip multiplexer selects the internal signals of interest for such monitoring. The number of test outputs is limited by the number of available bondpads (e.g., 8–16). As the digital signals internal to the ADPLL are clocked at the FREF rate (e.g., 100 MHz), they should be output at the same rate or down-sampled synchronously to ensure signal integrity. Propagating the 8-bit digital signals at the 100-MHz rate requires attention to the test circuit board design and the use of properly shielded test cables to minimize crosstalk. Moreover, these parallel outputs toggle between 0 (i.e., ground) and 1 (i.e., digital supply voltage), generating switching transients that could couple to the sensitive analog nodes on-chip via bondwires or the ESD/pad ring. Consequently, a higher noise floor and increased spurious tone levels are typically measured at the RF output once the digital test outputs are enabled. In addition, most of the critical internal signals discussed above are more than 8 bits in length to suppress quantization noise for sufficient data accuracy.

Table 8.1 Major trigger events for system snapshotting

1	System start
2	LF integral path enable
3	Dynamically change PLL loop BW
4	IIR filter enable
5	DCO gain calibration start/finish
6	TDC gain calibration start/finish
7	Loop locks (Φ_E fall in predefined window)
8	PLL loses lock
9	Modulation start
10	DCO tuning word overflow
11	Phase error frozen enable
12	TDC fail flag on

Therefore, a 32-bit signal has to be output in four successive clock cycles and reconstructed in a logic analyzer. Furthermore, it is impossible to monitor multiple internal signals simultaneously, which otherwise could be very powerful in some cases (e.g., monitoring R_v and ε to observe misalignment in the fractional and integer paths of the variable phase).

To overcome the aforementioned limitations of the parallel test outputs, we make use of on-chip SRAM to take a system snapshot that records one or more internal signals in the time frame of interest. Subsequently, the saved data are read out from SRAM via a serial interface (e.g., SPI) and displayed using a graphical user interface (GUI) in Matlab™, as indicated in Figure 8.1. Sharing the on-chip SRAM that is used for other (non-conflicting) purposes in an SoC reduces hardware overhead. Consequently, the digital signals and the timeframe recorded are selected carefully in order to utilize the limited word depth of SRAMs, and to fulfill debugging needs.

The important digital signals for monitoring in various scenarios were discussed in the previous section. A series of events, when the loop is prone to disruptions or even bugs, is defined (listed in Table 8.1) to specify the proper timeframe for recording, and to trigger the snapshot. One (or several) trigger events are selected for monitoring by a special control register.

Some of the events in Table 8.1 indicate a loop status change and can be enabled intentionally during debugging, such as switching the loop from type I to type II, increasing/decreasing the loop bandwidth, or enabling a higher-order IIR filter. Some indicate a different loop operation mode, for example, enabling DCO/TDC gain calibration, enabling the phase error glitch removal circuit, freezing the DCO coarse-tuning bank

(CB), starting the modulation, etc. The remaining events are internal flags (i.e., read only). For example, when no transition edges at the outputs of the TDC inverter chain are detected, the TDC fail flag is set to 1. When the DCO tuning word exceeds its range, an overflow flag turns on. Once the phase error variation exceeds a predefined window, the poor-clock-quality flag is activated. These flag signals report an abnormal status during the PLL operation. With the aid of system snapshots triggered by these flags, we are able to not only discover the loop abnormality, but also to infer the cause by examining associated digital signals.

Once the event of interest is triggered (externally/internally), the values of the associated digital signals (can be one, or several) are written into SRAM at a programmable clock rate (f_{clk_w}). Rate f_{clk_w} normally equals the clock rate of these digital signals (i.e., f_R *for* $\Phi_E[k]$) in order to capture their precise trajectory. In some cases, a down-sampled digital signal stream could be sufficient. For example, the filtered Φ_E (after the IIR filter) is a slowly varying signal whose trend can be obtained by sampling it at a much lower clock rate (e.g., $f_R/16$). Consequently, a longer time window can be captured with little sacrifice in accuracy of the recorded data. This provides great flexibility to meet varied testing requirements with limited on-chip storage. In this design, 16 critical internal signals can be monitored. Six SRAMs of 2^{13} bits each are used for multi-GHz FMCW generation, and are reused for ADPLL debugging. Clock f_{clk_w} is programmable from f_R and $f_R/2$ down to $f_R/16$.

To make the system snapshot even more powerful, we can configure it into different operation modes. *Mode1* saves one digital signal into all SRAMs in sequence to maximize the snapshot depth. *Mode2* can save up to six different signals into six small SRAMs in parallel to observe these multiple signals synchronously. *Mode3* stops saving data when the SRAMs are full, in order to capture the short moment when the trigger is enabled. *Mode4* saves the data to SRAM cyclically until triggered by the specified event to freeze the moment just before the event happens. In this work, a full precision of Φ_E (32 bit) trajectory up to 40 μs long can be captured (sampled at f_R) when all six SRAMs are used, which is quite sufficient for debugging purposes.

Figure 8.5 shows a snapshot of Φ_E samples at the LF output (i.e., NTW in Figure 8.3) triggered by a loop BW change. It captures the transient behavior of the PLL when the loop BW is decreased dynamically from 1.2 MHz to 300 kHz. The rms value of the NTW is directly proportional to the loop BW, and is thus reduced by a factor of 4 (as expected), which implies improved integrated PN at the RF output.

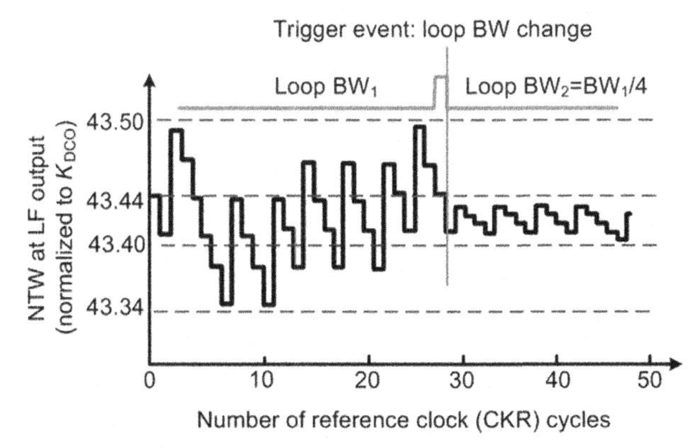

Figure 8.5 Filtered digital phase error samples at LF output for a dynamic loop bandwidth change captured by a combination of mode 3 and 4.

8.3.2 Performance-Based BIST and BISC

Two categories of tests are normally involved in an RFIC: structural-based and functional performance-based. While the former is used for block-level design verification and defect tests, the latter is used for PVT characterization.

8.3.2.1 DCO Tuning Step Analyzer

The digitally controlled oscillator is tuned by an oscillator command word (i.e., OTW in Figure 8.2). A simplified mm-wave DCO schematic is shown in Figure 7.1(a). It consists of three capacitor tuning banks of progressively finer step-size to simultaneously realize a wide tuning range and fine resolution [10]. The switched-capacitor array represents a digital-to-analog conversion function. Mismatches cause distortions in the DCO's modulation and, to a lesser extenf, frequency tracking.

For the 60-GHz FMCW transmitter design example, frequency modulation traverses three tuning banks (i.e., coarse-, mid-coarse-, and fine-bank) to obtain the desired GHz modulation range (see Figure 7.1b). The tuning step size (i.e., DCO gain, K_{DCO}) mismatch within each unit-weighted bank, and the K_{DCO} ratio between different banks affects the transmit spectrum emissions and modulation distortion. It is important to establish the tolerable extent for these mismatches in the design phase, and to verify that it is not exceeded by the fabricated SoCs.

It should be emphasized that it is extremely difficult and time-consuming to measure tuning step mismatches of a free-running, mm-wave

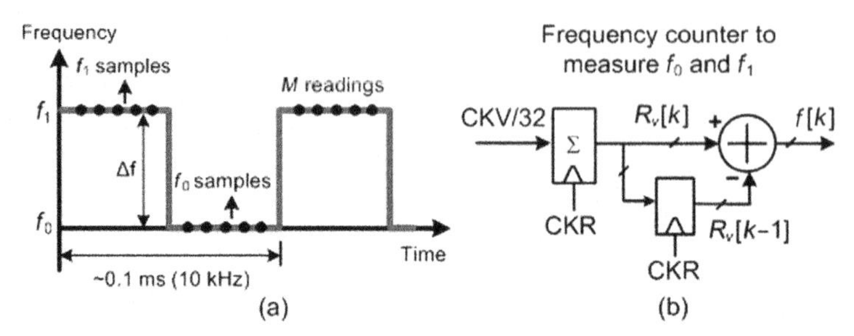

Figure 8.6 DCO tuning step analyzer: (a) DCO output frequency when toggled by a single bit and (b) on-chip frequency counter.

(e.g., 60 GHz) DCO for the targeted accuracy, even with the aid of specialized test equipment. For a fine-tuning bank (FB) K_{DCO} of ~1 MHz and 5% mismatch, the frequency difference is just 50 kHz at the 60-GHz carrier. Tremendous efforts would be required to stabilize the free-running DCO and to reduce thermal noise in the test setup (e.g., noise of a harmonic mixer should be avoided).

Fortunately, the BISC technique extracts the tuning characteristic of the DCO accurately without external resources. This BISC technique is similar to the method employed to characterize mismatch of the $\Sigma\Delta$ dithering bit in FB, with respect to the average K_{DCO} of the integer bits in FB, as illustrated in Section 7.3. The K_{DCO} for each tuning bit is characterized open-loop by the forced on/off toggling of individual digital bits. Consequently, the DCO output is toggled between f_1 (control bit on) and f_0 (control bit off), as shown in Figure 8.6a. The K_{DCO} for this control bit is simply Δf, which is evaluated by subtracting frequency measurements performed at each of the two states. Frequency measurements are made using a counter within the ADPLL, as shown in Figure 8.6b. The frequency is measured by counting the variable phase change in one FREF cycle (i.e., $f[k] = Rv[k] - Rv[k-1]$). Multiple readings of the counter (e.g., M readings) are averaged out to reduce the quantization error of a single frequency measurement, especially in the presence of DCO phase noise. The tuning step (f_i) for a particular control bit "i" is calculated by Eq. (7.1). The frequency tuning step after averaging is shown in Eq. (7.2). Thus, the normalized tuning step mismatch of this bit is $e = (\Delta f_{avg,i} - \Delta f_{FB})/\Delta f_{FB}$, where Δf_{FB} is the average K_{DCO} of the FB.

BISC results are compared to the simulation data for a block-level design verification. In addition, they can be compared to statistically chosen thresholds for defect detection. Furthermore, a "self-healing" capability can be implemented based on the BISC outcomes. For example, an on-chip lookup table can be built based on the BISC results, and used to pre-distort the modulation data in order to compensate the mismatch and retrieve adequate linearity [11].

8.3.2.2 Built-in Phase Error (Φ_E) Analyzer

As explained in Section 8.2, the performance at the RF output can be determined with adequate accuracy by observing the internal digital phase error signal Φ_E, which is sampled at the much lower rate of f_R. The filtered output of the phase detector (or its raw version) is sufficient to indicate several parameters of interest in the RF output, such as frequency error, phase noise at lower-frequency offsets, integrated phase noise, and rms phase trajectory error upon modulation. System snapshotting described in Section 8.3.1 stores the Φ_E trajectory into the on-chip SRAM for a timeframe of interest, providing valuable data when identifying bugs. For performance testing of mass-produced ICs, on-chip digital signal processing Φ_E is desirable, which generates flags that characterize the performance of the RF output.

A simple phase error analyzer is implemented in the 60-GHz ADPLL IC, as shown in Figure 8.7. The raw phase error Φ_E is compared with its sample at the previous clock cycle, which represents the phase error change ($\Delta\Phi_E$) in one FREF cycle, or the frequency error (f_E) representing an instantaneous frequency drift. Once the PLL is locked, f_E is zero

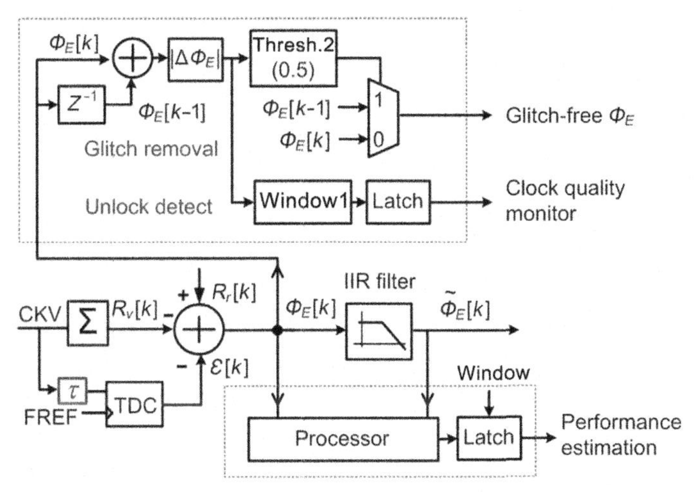

Figure 8.7 Phase error analyzer.

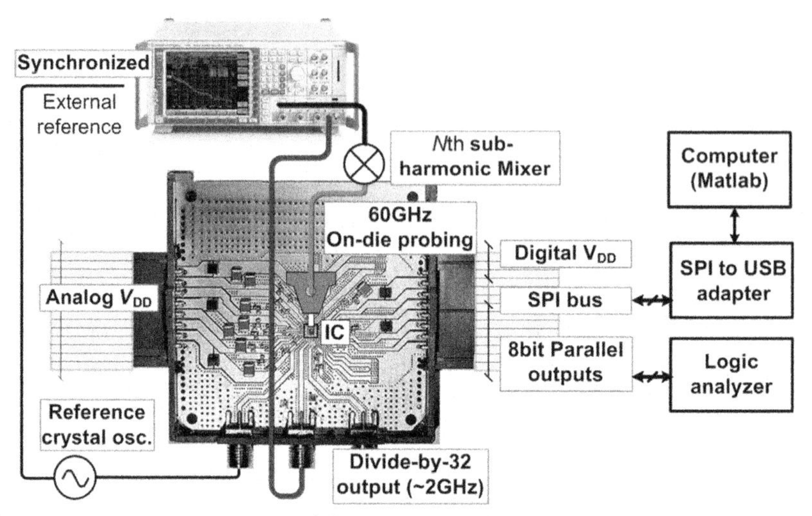

Figure 8.8 Measurement setup of the 60-GHz ADPLL-based transmitter.

with some small quantization noise. Thus, the quality of the synthesized clock can be monitored by constantly comparing f_E with a programmable threshold. If f_E exceeds a large threshold for a number of FREF cycles (controlled by the counter in Figure 8.7), the loop has likely lost its lock, since a large frequency error due to an unwanted incidental disturbance would normally not have lasted that long. A simple clock quality monitor is thus realized.

The phase error analyzer can be further extended to estimate the PLL's in-band PN, the DCO's PN (at a narrow loop BW configuration, e.g., 10 kHz), and modulation noise in a transmitter with the aid of an on-chip stream processor [5,6].

8.4 MEASUREMENT SETUP AND PROCEDURES

8.4.1 Measurement Setup

The DFT and DFC techniques discussed in this chapter have been realized as part of the 60-GHz ADPLL-based FMCW transmitter implemented in 65-nm bulk CMOS. The chip micrograph is shown in Figure 6.25. The ADPLL core occupies 0.5 mm^2 of the 2.2-mm^2 total die area, including bondpads, power amplifier, SRAMs (6 × 8-kbit), and other digital circuitry for debugging. SRAMs store the DCO gain calibration data for the GHz range, linear FM sweeping, and are reused to take system snapshots during debugging.

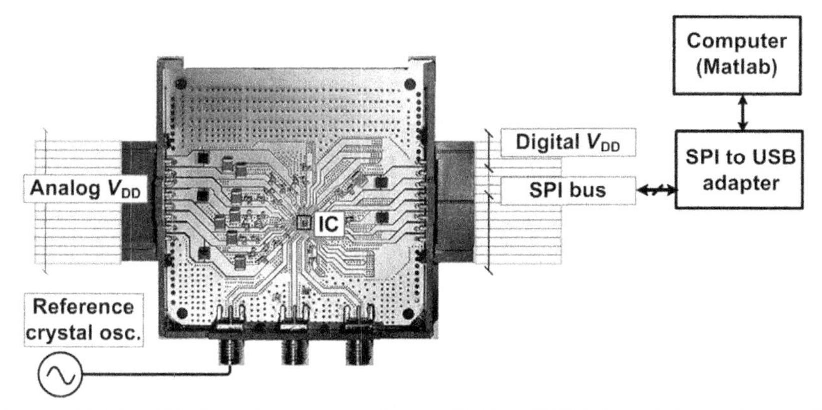

Figure 8.9 Simplified testing setup of the 60-GHz ADPLL-based transmitter with BIST.

The measurement setup is depicted in Figure 8.8. The IC is wire-bonded to a PCB for DC and low-frequency connectivity, while the 60-GHz PA output is probed on-die. A divide-by-32 test output provides a convenient way to monitor the 60-GHz ADPLL output without the on-die probing. An on-chip 4-bit serial peripheral interface (SPI) controls 256 internal (write/read) registers and the SRAMs, providing reconfigurability and flexibility for both open-loop and closed-loop tests. The 8-bit digital test outputs are captured in parallel via a logic analyzer. With the BIST and BISC techniques discussed in this chapter, the measurement setup can be simplified to Figure 8.9, in which no external RF equipment is employed. The self-testing results can be read out via the SPI interface.

Matlab™ scripts are developed to make the ADPLL testing automatic or semi-automatic. The graphic user interfaces (GUI) shown in Figure 8.10 are designed to visualize the internal register status and facilitate the testing process.

8.4.2 Test Procedures and Test Modes

The 60-GHz ADPLL IC supports three major test modes: open-loop test, closed-loop CW-mode test, and the frequency modulation test. The key testing cases/steps are summarized in Table 8.2. The testing procedures are as follows.

First, SPI self-check and SPI access to the SRAMs are performed to verify the communication between the external control software (Matlab™) and the ADPLL chip. This is the only way to access the internal registers

(a)

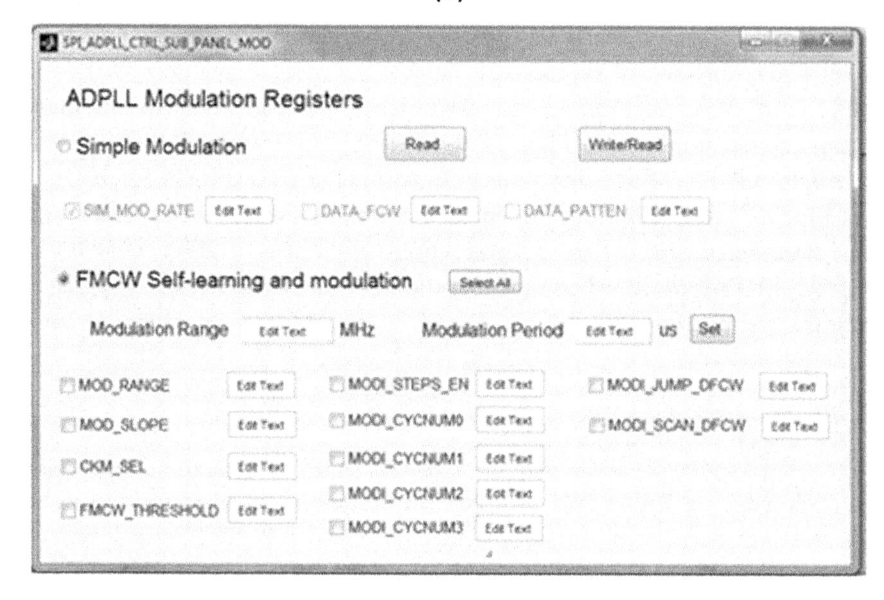

(b)

Figure 8.10 GUI interface for ADPLL testing: (a) main GUI and (b) for modulation control.

Table 8.2 Major test cases for the 60-GHz ADPLL system

Test cases	**Remarks**
Step 1.1	SPI self-check (R/W 128-byte registers)
Step 1.2	SPI access SRAM (R/W)
Step 1.3.1	Open-loop test: DCO stand alone (divider disabled)
Step 1.3.2	Open-loop test: DCO with divider loading
Step 1.4	Open-loop test: Divider chain
Step 1.5	Open-loop test: FREF slicer
Step 1.6	Open-loop test: TDC functionality test
Step 2.1	DCO open-loop mismatch characterization
Step 3.1	ADPLL closed-loop test: lock to a single frequency
Step 3.2	ADPLL closed-loop test: DCO gain calibration
Step 3.3	ADPLL closed-loop test: TDC gain calibration
Step 3.4	ADPLL closed-loop test: externally control LF via SPI
Step 3.5	ADPLL closed-loop test: 1st/2nd $\sum\Delta$ modulation for DCO
Step 4.1	Simple FSK modulation
Step 5.1	FMCW initialization: multi-bank DCO gain calibration
Step 5.2	FMCW modulation
Step 6.1	Trigger event and system debug
Step 7.1	Other function test: phase error freeze
Step 7.2	Other function test: integer-mode operation (disable TDC)

in the ADPLL prototype, with which to configure both analog and digital parts. Second, a number of open-loop tests are conducted: DCO standalone test, DCO test with divider loading, divider chain test to verify the variable clock path; reference slicer and TDC checks to ensure that the reference clock generation and FREF retiming works properly. Up to now, the low-speed digital part of the loop is unclocked and remains quiet. Therefore, the measured performance of the RF/analog building blocks is not affected by the digital activity. The success of the above open-loop checks ensures generation of the synchronous system clock (CKR) for the low-speed digital part and enables the closed-loop test.

The tuning characteristics of the DCO are then measured. The 60-GHz DCO employs three switched-metal capacitor banks distributed across a transformer-coupled resonator. The K_{DCO} for each bank measured via the BISC technique described in Section 8.3 is plotted in Figures 8.11–8.14. The measured frequency resolution (K_{DCO}) of the CB for each thermometer code bit is shown in Figure 8.11 for five IC samples. The CB achieves an average K_{DCO} of 400 MHz/bit and an in-band mismatch of 15%, employing the digitally controlled TL. Figure 8.12 plots the measured K_{DCO} of the mid-coarse tuning bank (MB) under different CB settings.

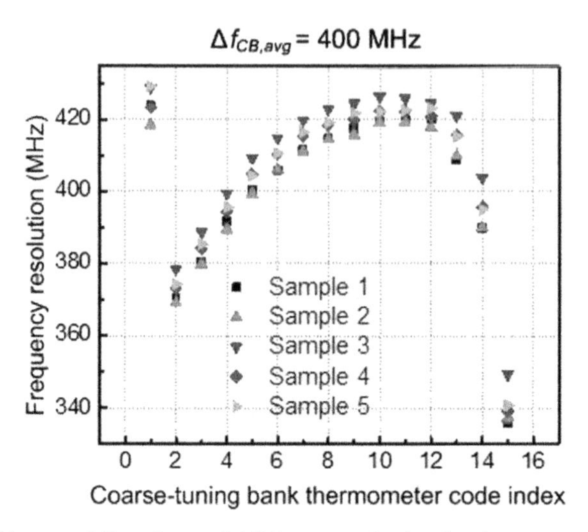

Figure 8.11 Measured K_{DCO} for each bit in coarse-tuning bank.

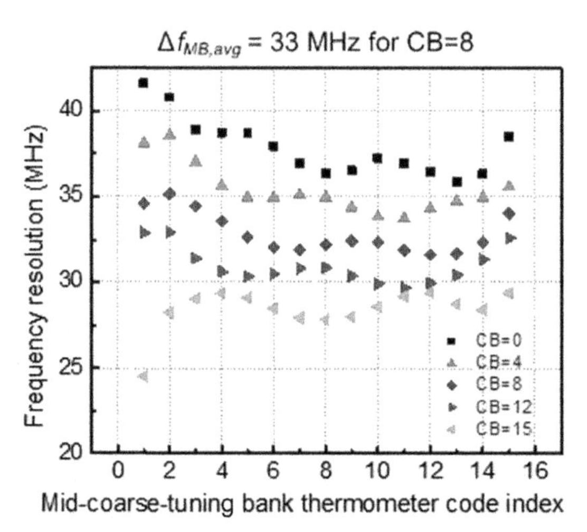

Figure 8.12 Measured K_{DCO} for each bit in MB for various CB settings.

When CB = 0, the DCO oscillates at its maximum frequency, and the K_{DCO} of MB is 38 MHz/bit (mean value) with a 15% mismatch. When CB increases, the oscillation frequency drops, and thus the K_{DCO} of MB decreases accordingly. It is clear that the K_{DCO} of each MB bit for various CB settings must be calibrated and compensated upon modulation in order to generate an ultralinear chirp of a gigahertz-level modulation range.

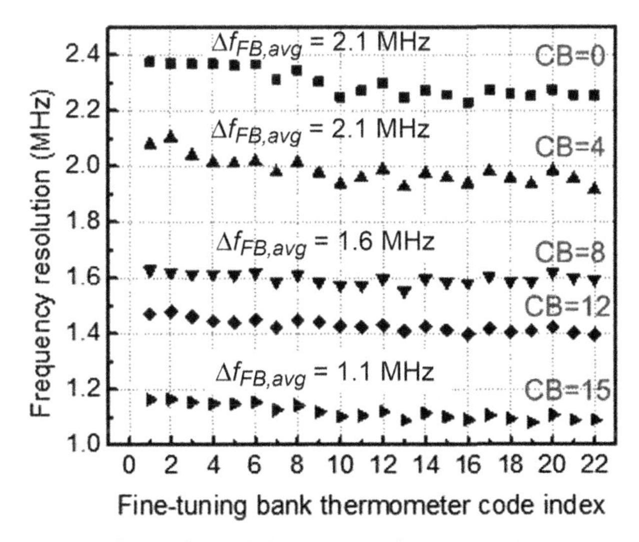

Figure 8.13 Measured K_{DCO} for each bit in FB_{Mod} for various CB settings.

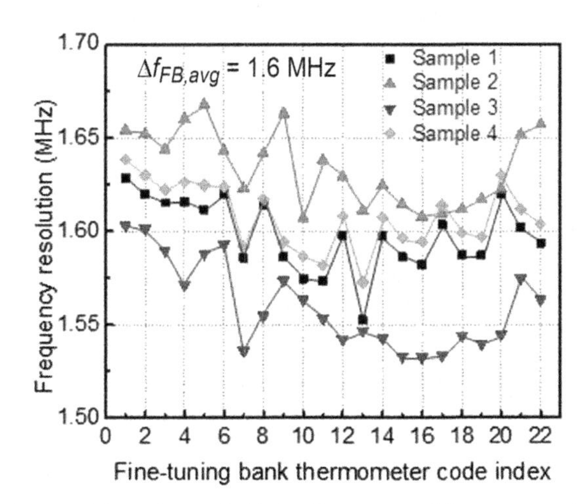

Figure 8.14 Measured K_{DCO} for each bit in FB_{Mod} over different samples (CB = 8).

The same trend (i.e., a drop in K_{DCO} with an increasing CB value) is observed for the FB and plotted in Figure 8.13. The digital-to-frequency conversion linearity of the FB measured for four IC samples is shown in Figure 8.14. The measurement accuracy achieved for the FB tuning step is 1 kHz via the aforementioned on-chip averaging.

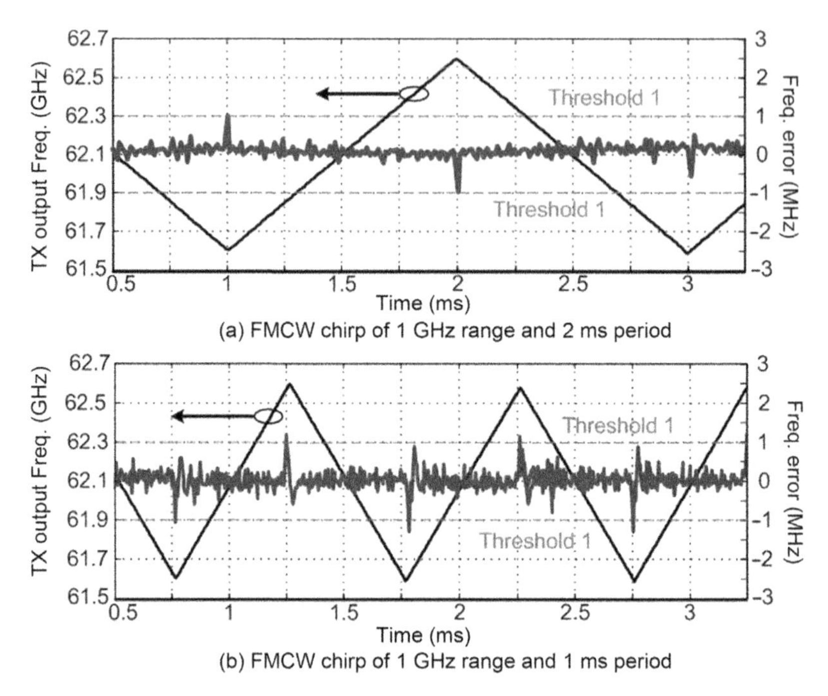

Figure 8.15 Measured transmitter output frequency and its error from ideal for an FMCW chirp.

In the closed-loop test, a system snapshot and the parallel digital test outputs are used to monitor important internal signals for troubleshooting. Moreover, a TDC self-test is conducted via the built-in TDC gain (i.e., TDC resolution) calibration algorithm described in Section 6.5.3. The loop filter and sequencer can be reconfigured via the SPI bus to observe the PLL performance under various loop bandwidths and the gearshift events. More closed-loop tests are listed in Table 8.2.

The third mode is the frequency modulation test. The measured FMCW transmitter performance is depicted in Figure 8.15 for a 1-GHz modulation range centered at 62.1 GHz with varied chirp slopes. The transmitter output frequency is measured internally via the frequency counter at the divide-by-32 output. It is evaluated against the ideal FM trajectory to obtain the instantaneous frequency error (plotted in Figure 8.15). The FMCW transmitter performance is determined using the built-in phase/frequency error analyzer, and without RF probing of the 60-GHz output. Moreover, the rms frequency error can be obtained with the aid of a stream processor. Both the rms and instantaneous frequency error could be further evaluated against statistically chosen thresholds (e.g., threshold 1 in Figure 8.15) in production testing.

8.5 SUMMARY

DFT and DFC techniques for mm-wave digitally intensive PLLs and frequency modulator were elaborated in this chapter, using a 60-GHz ADPLL-based FMCW transmitter as the demonstrator. The digitally intensive architecture facilitates BIST and BISC with little hardware overhead, by monitoring and storing internal digital signals in SRAM, and with the aid of digital signal processing. Snapshotting key internal signals captures the loop transient behavior for the timeframe of interest, providing valuable data for debugging. Phase error and DCO tuning step analyzers can be used to characterize the RF performance without employing external resources. Measured key parameters can be evaluated against statistically chosen performance measures in production testing. Finally, these techniques could also be applied to other digitally intensive PLLs to reduce test time and cost.

REFERENCES

[1] S. Kim, M. Soma, An all-digital built-in self-test for high-speed phase-locked loops, IEEE Trans. Circuits Syst. II 48 (2) (2001) 141–150.

[2] M. Burns, G.W. Roberts, An Introduction to Mixed Signal IC Test and Measurement, Oxford University Press, 2001.

[3] P. Goteti, G. Devarayanadurg, M. Soma, DFT for embedded charge-pump PLL systems incorporating IEEE 1149.1, in: Proceedings of IEEE Custom Integrated Circuits Conference, May 1997, pp. 210–213.

[4] M.F. Toner, G.W. Roberts, On the practical implementation of mixed analog–digital BIST, in: Proceedings of IEEE Custom Integrated Circuits Conference, May 1995, pp. 525–528.

[5] R.B. Staszewski, I. Bashir, O. Eliezer, RF built-in self test of a wireless transmitter, IEEE Trans. Circuits Syst. II 54 (2) (2007) 186–190.

[6] O. Eliezer, O. Friedman, R.B. Staszewski, A built-in tester for modulation noise in a wireless transmitter, in: Proceedings of Fifth IEEE Dallas Circuits and Systems Workshop: Design, Application, Integration and Software (DCAS-06), October 2006, pp. 59–62.

[7] O.E. Eliezer, R.B. Staszewski, J. Mehta, et al., Accurate self-characterization of mismatches in a capacitor array of a digitally-controlled oscillator, in Proceedings of IEEE Dallas Circuits and Systems Workshop, October 2010, pp. 1–4.

[8] O.E. Eliezer, R.B. Staszewski, Built-in measurements in low-cost digital-RF transceivers, IEICE Trans. Electron. E94-C (6) (2011) 930–937.
[9] H. Sakurai, Y. Kobayashi, T. Mitomo, et al., A 1.5 GHz-modulation-range 10 Ms-modulation-period 180 kHz rms-frequency-error 26 MHz-reference mixed-mode FMCW synthesizer for mm-wave radar application, in IEEE International Solid-State Circuits Conference Digest of Techical Papers, February 2011, pp. 292–293.
[10] W. Wu, J.R. Long, R.B. Staszewski, High-resolution millimeter-wave digitally-controlled oscillators with reconfigurable passive resonators, IEEE J. Solid-State Circuits 48 (11) (2013) 2785–2794.
[11] W. Wu, R.B. Staszewski, J.R. Long, A 56.4-to-63.4 GHz multi-rate all-digital fractional-N PLL for FMCW radar applications in 65-nm CMOS, IEEE J. Solid-State Circuits 49 (5) (2014) 1–16.

INDEX

Note: Page numbers followed by "*f*" and "*t*" refer to figures and tables, respectively.

B

C

D